Nuclear Power
Past, present and future

Nuclear Power

Past, present and future

David Elliott
Open University, UK

A 1950s UK miss: Harwell researchers and the ZETA Zero Energy Thermonuclear Assembly. It was claimed that they had achieved fusion. Sadly, it proved not to be the case.

Morgan & Claypool Publishers

Copyright © 2017 Morgan & Claypool Publishers

All rights reserved. No part of this publication may be reproduced, stored in a retrieval system or transmitted in any form or by any means, electronic, mechanical, photocopying, recording or otherwise, without the prior permission of the publisher, or as expressly permitted by law or under terms agreed with the appropriate rights organization. Multiple copying is permitted in accordance with the terms of licences issued by the Copyright Licensing Agency, the Copyright Clearance Centre and other reproduction rights organisations.

Rights & Permissions
To obtain permission to re-use copyrighted material from Morgan & Claypool Publishers, please contact info@morganclaypool.com.

ISBN   978-1-6817-4505-3 (ebook)
ISBN   978-1-6817-4504-6 (print)
ISBN   978-1-6817-4507-7 (mobi)

DOI   10.1088/978-1-6817-4505-3

Version: 20170401

IOP Concise Physics
ISSN 2053-2571 (online)
ISSN 2054-7307 (print)

A Morgan & Claypool publication as part of IOP Concise Physics
Published by Morgan & Claypool Publishers, 40 Oak Drive, San Rafael, CA, 94903 USA

IOP Publishing, Temple Circus, Temple Way, Bristol BS1 6HG, UK

Contents

Preface	vi
Guide to abbreviations, terms, units and a note on bias	vii
Acknowledgments	ix
About the author	x

1 Introduction: the nuclear vision — 1-1

1.1	Nuclear energy: uranium in a bucket	1-1
1.2	Atoms for peace	1-3
1.3	The rise and fall of nuclear power	1-6
1.4	On to Generation IV?	1-10
	References	1-12

2 Nuclear innovation: the early days — 2-1

2.1	Early US experiments	2-1
2.2	Thorium reactors and fast breeders	2-3
2.3	Generation IV design parameters and constraints	2-9
	References	2-12

3 New brooms: on to Generation IV — 3-1

3.1	Back to breeders, and thorium	3-1
3.2	Small is beautiful	3-7
3.3	Reactor choices and progress	3-11
	References	3-15

4 Nuclear power revisited — 4-1

4.1	A review of the prospects for new nuclear	4-1
4.2	What long-term future for nuclear?	4-7
4.3	Conclusions: the way ahead	4-11
	Afterword: insider views	4-16
	Further reading/videos	4-18
	References	4-19

Appendix: Nuclear and renewables—the basics compared — A-1

Preface

Nuclear fission produces heat, which has been used for electricity generation in a range of power plants, with a varying degree of success. Early hopes were that nuclear power would become a major energy source. However, the initial quite rapid expansion in the 1960s and 70s was slowed by increasing competition from cheaper alternatives, including natural gas, and by major nuclear disasters. An attempt at a nuclear renaissance, based in part on improved Generation II reactor designs and the argument that nuclear power could play a role in responding to climate change, was constrained by the Fukushima disaster in 2011 and increased competition from other sources. The result is that nuclear has in effect stalled at around a 11% global electricity share, while, by 2016, newly emergent renewables had reached 24%, and seem likely to dominate in the years ahead, with their costs falling.

Although some nuclear energy enthusiasts hope that better progress will be made with the current upgraded Generation II designs, the so-called Generation III reactors, there have been continuing technical, economic and project delivery problems, and some say the future for nuclear, if there is to be one, will lie in developing a new set of technologies, Generation IV. However, many of these options are not in fact new–they were looked at in the early pioneering days of nuclear power development, and in most cases abandoned. There were some dramatic failures, but many of the basic ideas that are now being re-explored were developed, ranging from liquid sodium cooled fast-neutron plutonium breeders to molten salt thorium fueled reactors, including some small-scaled plants.

The context is now very different, with tighter economic, environmental, safety and security constraints and renewables are also presenting a major challenge, but some nuclear enthusiasts are hopeful that updated variants of these early ideas can win through, offering higher fuel efficiency and reduced waste production. However, if some of them are to play a major role in future, they will have to overcome whatever problems led to their demise then, and also the problems that have faced subsequent designs. This book looks at the early history of nuclear power, at what happened next and at its longer-term prospects.

Guide to abbreviations, terms, units and a note on bias

Some major reactor types

BWR	boiling water reactor
EPR	European pressurised-water reactor
HTR	high temperature reactor
IFR	integral fast reactor
LFTR	liquid fluoride thorium reactor
MSR	molten salt reactor
PWR	pressurised water reactor
SMR	small modular reactor

LWR   light water reactor, generic term for reactors cooled/moderated with ordinary water as opposed to *heavy* water, i.e. with deuterium rather than hydrogen atoms

A quick guide to key atoms...

Some can be made to fission (spilt) and so are said to be 'fissile'.

U235	a rare fissile isotope of uranium
U238	what most uranium consists of: it is 'fertile', i.e. it can be converted to:
Pu239	plutonium, a fissile material, made from U238 by absorbing a neutron
Th232	thorium is not fissile, but is fertile, i.e. it can be converted to:
U233	a fissile product of thorium, made from Th232 by absorbing a neutron

Neutrons are electrically neutral particles that exist (along with protons) in the nucleus of most atoms and which can be emitted when the atom disintegrates, e.g. as a result of fission.

Energy units

Reactor **power** generation capacities are cited in megawatts (MW) or gigawatts (GW): 1 GW is 1000 MW, 1 MW is 1000 kilowatts (kW). A typical one bar electric fire power rating will be 1 kW. Running it for an hour would use 1 kilowatt hours (kWh) of **energy**. A 1 MW rated wind turbine, if it could be run flat out, would generate 1000 kWh, or 1 MWh, a 1 GW nuclear plant 1000 MWh, or 1 GWh, but energy outputs are usually quoted in *annual* terms.

In what follows, to make it accessible to the non-specialist reader, and for brevity, the physics has been kept to a minimum and much simplified, some might say too much.

Bias

The nuclear power field is contentious, with strong pro- and anti-nuclear views often being expressed. This study strives to avoid polarised extremes, but perhaps inevitably, since it adopts a critical approach, it may be seen as biased by nuclear enthusiasts. Though it may also be seen as not critical enough by those strongly opposed to nuclear power.

Acknowledgments

Title page photo: United Kingdom Atomic Energy Authority, used with permission.

Figure 1.1 © OECD/IEA 2011 World Energy Outlook, IEA Publishing. Licence: www.iea.org/t&c.

Figure 2.1 © OECD/IEA 2007 Energy R&D Statistics Database http://www.iea.org/Textbase/stats/rd.asp IEA Publishing. Licence: http://www.iea.org/t&c/.

Figure 2.2 from the World Nuclear Industry Status Report, reproduced with permission.

Figure 3.1 from GIF 2014 report published by the OECD, available for free download from the GIF portal https://www.gen-4.org/.

Figure 4.1 from BNEF data, reproduced with permission.

About the author

David Elliott

David Elliott worked initially with the UK Atomic Energy Authority at Harwell and the Central Electricity Generating Board before moving to The Open University, where he is now an Emeritus Professor. During his time at The Open University he created several courses in design and innovation, with special emphasis on how the innovation development process can be directed towards sustainable technologies. He has published numerous books, reports and papers, especially in the area of developing sustainable and renewable energy technologies and systems.

IOP Concise Physics

Nuclear Power
Past, present and future
David Elliott

Chapter 1

Introduction: the nuclear vision

Nuclear power was once seen as the ultimate energy source, freeing mankind from reliance on dirty and expensive fossil energy. In the early 1950s there were utopian visions of energy plenty and unbounded prosperity fuelled by cheap nuclear electricity and ever more rapid technological advance around the world. Contact with reality has reduced some of this enthusiasm. Sixty years on, nuclear only supplies around 11.5% of global electricity (WNA 2017). There are still optimistic projections about the prospects for new nuclear power, but, in the main, it has been the new entrant, renewable energy, that has seen most promise and most growth, wind and solar especially. These sources, along with hydro, now supply around 24% of global electricity (REN21 2016). There are projections that it could continue to expand to supply 50% by 2030, and maybe even almost all electricity needs globally by around 2050. Some of these projections may be overblown, much as were the early projections for nuclear power. However since, once again, we are being offered some inspiring visions of what technology can do, it may perhaps be helpful to look at what went wrong (and right) last time, with nuclear power. That may help us to put renewables in perspective and also to make a better, more robust, assessment of what role nuclear (a non-renewable resource) might play in the future.

1.1 Nuclear energy: uranium in a bucket

Nuclear energy developed out of the Manhattan nuclear bomb programme in the US in the 1940s. The **fission**, or splitting, of uranium 235 atoms produces heat and radiation, including neutrons. Some of the neutrons collide with other 'fissile' U235 atoms and, under the right conditions, that can lead to a runaway chain reaction, many more fissions and an explosion. If some of the neutrons are slowed by a so-called *moderator* (water is one option, graphite another), the reaction can be controlled, so that the reaction provides continuous heat, which can raise steam to drive a turbine to generate electricity. However, the reaction also produces other things. U235 is a rare isotope of uranium—most uranium consists of non-fissile U238. Depending on the level of U235 enrichment (i.e. its concentration), reactor fuel is mostly U238, some of which is unavoidably converted, by absorbing neutrons, into a new type of atom, plutonium, Pu239. That is fissile, like U235, and can be used as a reactor fuel. It is also what is mostly used in atom bombs (though highly enriched U235 can also be used), the Pu239 being extracted from used reactor fuel via so-called *reprocessing* operations. What is left after that, is a residue of highly radioactive waste, along with any unconverted U238/235.

Civil nuclear reactors thus inevitably produce plutonium and wastes, but they are designed mainly for heat production. Heat is extracted by pumping gas, or more usually water, through the reactor core, although in plutonium fueled 'fast neutron' reactors, liquid metals are used, sodium or molten lead, as coolant, in a smaller, more compact core. In this configuration, the neutrons are not slowed as in conventional, so called 'thermal', slow-neutron reactors, i.e. they are 'fast', higher energy, neutrons. They can however cause fission in plutonium, and that releases more neutrons than uranium fission, which can then go on to be absorbed by U238, as well as causing more Pu fissions. So 'fast reactors', running mostly on plutonium, are good at converting U238 into more Pu239, which also then undergoes fissions, sustaining a plutonium 'breeding' process. Hence the term 'fast breeder', although the breeding rate is not actually fast. It can take time to produce a replacement amount of plutonium and the initial Pu239 charge still has to be produced in a thermal uranium reactor.

Most of the reactors in use at present are simple 'thermal neutron' uranium-fuelled reactors, although some use a mix of uranium and plutonium oxide (MOX). The use of other types of fuel is possible, including thorium 232, reserves of which are around three times more abundant than uranium. However, thorium 232 is not fissile, and has to be converted into a form (U233) that is, using a source of neutrons, obtained via the fission of uranium or plutonium, or by using a particle accelerator to create a neutron beam.

At very high temperatures (millions of degrees) isotopes of hydrogen atoms can be fused, and this **fusion** reaction releases heat and radiation. It is what happens in the Sun. So far, we have been unable to sustain fusion reactions on Earth for more than a few seconds, but in theory it should be possible to build fusion reactors with sustained energy output and use it to generate power. However, that is some way off. As the overhyped ZETA prototype fusion project in the UK demonstrated in the 1950s, it is sometimes best not to be too optimistic (Pease 2008).

Fusion work continues but, so far, most of the emphasis for nuclear plants around the world has been on fission, and on what, to put it simply, is steam production from hot uranium in a bucket. Portrayed that way, it sounds straight-forward, but clearly it has taxed some of the best scientists and engineers. As will be seen from the accounts below, it has not always been easy or without drama, but, for protagonists, belief remained strong that nuclear technology, despite being developed initially for war, could become a major boon for humanity.

1.2 Atoms for peace

In a December 1953 speech to the United Nations, President Dwight D Eisenhower launched what was soon to labeled the 'Atoms for Peace' programme. Its motivations were varied. It was ostensibly an attempt to show that, as he put it, *'the miraculous inventiveness of man shall not be dedicated to his death but consecrated to his life'*. The horrors of the atom bomb, as visited by the US on Japan in 1945 in wartime, were now to be balanced by the wonders of civil nuclear power, with nuclear technology being turned to peaceful ends around the world, with US help. Moreover, Eisenhower claimed that *'peaceful power from atomic energy is not a dream of the future. That capability, already proved, is here—now—today'* (Eisenhower 1953).

However, in his haste to redeem the US and perhaps also ensure that it retained commercial and political control over how this technology was used and sold, he arguably overstated its readiness. No US nuclear plant had been built at that point, and internal US government reports soon pointed out that nuclear power was neither economic nor ready for export.

The Atoms for Peace propaganda offensive mounted by the US Information Agency (USIA) and others was certainly effective in selling the positive vision, but, with progress on the ground being very limited, those responsible for delivering it began to be concerned that the rhetoric being used was all too 'pie in the sky' in terms of what was available technologically and also what might be realistic in terms of applications. For example, as Mara Drogan notes in a wide-ranging review of the programme: *'An Atoms for Peace exhibit being prepared for India by the USIA highlighted nuclear power plants, but all studies indicated that India would not have the need or ability to install such plants for some time to come'*, while Gerard Smith, John Foster Dulles's Special Assistant for Atomic Energy, was worried about *'the impression created by these exhibits as to the imminence and benefits of nuclear power'*. He warned that *'we may lose good will rather than gain it if through these exhibits we arouse expectations which cannot be fulfilled in the reasonably near future'*. (Drogan 2016).

Drogan quotes from a classified State Department Intelligence Report, circulated in January 1954, 'Economic Implications of Nuclear Power in Foreign Countries', which noted that it was likely to be more expensive than conventional electricity generation: *'Nuclear power plants may cost twice as much to operate and as much as 50 percent more to build and equip than conventional thermal plants'*. So it warned that the introduction of nuclear power would *'not usher in a new era of plenty and rapid economic development as is commonly believed'*. It suggested that the only likely candidates for potentially economical nuclear power were countries such as Great Britain and Japan, which were already highly industrialised and had diminishing reserves of fossil fuels and limited options for expanding hydro-electricity. However, it added, although it would be too costly for the foreseeable future for developing countries, some might seek to develop it anyway, *'for reasons of national prestige or in the hope that technological developments and engineering experience will promote lower costs'*.

Looking back with hindsight, that was a fairly good set of predictions. The reality has been that nuclear only developed slowly and initially just in the West, but also in Russia, followed later by Japan, and then China, South Korea, India and Pakistan. Moreover, in at least some of these cases, and others that came after, the impetus has arguably been as much military as civil, with the distinction between them often being blurred. India evidently used a Canadian-supplied CANDU plant (which runs on non-enriched uranium) to get material for its bomb. Certainly not Atoms for Peace! The potential for 'dual use' has proved to be major issue for the expansion of nuclear power, as witness the long battles with Iran over its nuclear plans.

Even within those countries who were to become signatories to the UN Non-Proliferation Treaty, the distinction was not always clear. For example, the early plants built in the UK were dual-purpose, able to produce power and/or materials for weapons if needed. The haste with which nuclear programmes were carried out, also led to problems. The UK built two air-cooled uranium piles at Windscale in Cumbria, designed to produce plutonium for its bomb, but, given the perceived need to demonstrate that the UK was nuclear weapons capable, there was strong political pressure to speed up production, with short-cuts to safe operational procedures allegedly being adopted. That seems to have been one of the reasons why there was a major disaster in 1957 at Windscale, with the fuel catching light, leading to the emission of radioactive material into the environment. Measures were taken to put the fire out and limit exposure (see box 1.1), but it is now thought that about 240 people downwind may have developed cancers (Morelle 2007). However, the nuclear bomb programme went on.

Box 1.1. Nuclear accidents. *Key ones reviewed: there have been others* **(Guardian 2011).**

All technologies can go wrong, but when and if nuclear plants fail the impacts can be very severe and long-lived. Small leaks and mishaps happen regularly, but, although they may add cumulatively to the amount of radioactive material in the environment, they inevitably get less publicity than the big very visible disasters. Sometimes accidents are put down to human error. This can be a matter of interpretation. The Windscale fire in Cumbria, in 1957, was mentioned earlier. The design of the plant combined with the haste to produce material for a bomb seem to have been the cause (Dwyer 2007), with the result being venting of radioactive material, though its release into the air was fortunately reduced by a filter that had been put in the up-venting chimney as a precaution. Even so, with the fire burning for three days before heroic efforts by the plant staff to extinguish it succeeded, some still escaped, and, as an after-the-event precaution, milk from cows grazing in the area was poured away.

The accident at Three Mile Island in Pennsylvania was, it seems, due to a combination of mechanical faults and human error. Two automatic pump valves failed, and another one had been left open after a test by mistake. A hydrogen explosion was narrowly averted by releasing some pressure, but that led to the release of radioactive gas. Many thousands of local people self-evacuated, but exposure levels were said to have been low, with few if any health impacts being reported, although that is disputed.

> The Chernobyl disaster in Ukraine was a result of an ill-conceived safety test (see box 2.3). The plant suddenly overheated in an uncontrolled way, in part due to its basic design, resulting in an explosion and a radioactive cloud moving across Europe, as shown in a map-based animation (ISRN 2011).
>
> 40 or so people died soon after, but over 4000 subsequent thyroid cancers (mostly treatable) were reported amongst children who were in the area. A similar number of early deaths has been claimed for the people who, heroically, carried out post-explosion clean up, as well as more deaths and illness amongst residents in the region (IAEA 2006). Independent estimates for total likely deaths range up to 60 000 (Fairlie and Sumners 2006). The area is still unsafe, with an exclusion zone in force.
>
> By contrast, the Fukushima disaster in Japan in 2011 was due to what might be called an act of God, a major tsunami, which overwhelmed the nuclear complex, which was ill-designed to cope. Heroic efforts by staff avoided a much worse disaster, after waste tanks were disrupted, three of the plants were demolished by hydrogen explosions, and the reactor cores melted down (Elliott 2013). Some staff were killed in the explosions and some workers have received compensation for cancers they developed, but no public radiation-linked deaths have been reported so far: fortunately, most of the radiation cloud initially went out to sea. However, there were contamination hot spots and 160 000 or so residents were evacuated. Many have yet to return. The long-term clean up operation will take decades, with the cost, including compensation payments, put at around £140 billion. Some estimates put possible long-term radiation-related deaths in the thousands (Ten Hove and Jacobson 2012, Fairlie 2014).

The Calder Hall plant built alongside the Windscale piles was also available for plutonium production, and its design, with carbon dioxide gas cooling, was the basis for a series of civil plants, with uranium in Magnox alloy tubes. The UK continued with gas cooling in the follow up 'Advanced Gas-cooled Reactors' (AGRs), in contrast to the US, which went for water-cooling for its civil nuclear plants. They too had military roots. The US had initially focused on developing small reactors for submarines, the first being in the Nautilus in 1954. The reactor had to be compact, so water-cooling was used and that was to be the basis of subsequent scaled-up US civil reactor designs. The most successful were the pressurised water-cooled reactor (PWR) and the boiling water reactor (BWR), with enriched uranium. PWRs, and some BWRs, were then adopted around the world, most notably in France, which pushed ahead with a major civil programme, in parallel with its weapons programme.

The French PWR programme in the 1970s was technically very successful, eventually supplying around 75% of the country's electricity, with the mass rollout of a more or less identical standard design leading to economies of scale. Even so, the rapid expansion was economically challenging. In the US, expansion (at peak 2000 MW pa) was driven by private investment, along with some state support for ancillary facilities, with the cost in the main being passed on to consumers. However, in France, in what was an almost entirely public sector programme, the high capital cost was met by government borrowing on world finance markets. That allowed for electricity prices to be set low but meant that the government faced punishing repayment costs.

Water-cooled and moderated PWR and BWR designs, so-called Generation II reactors, have remained the mainstay around the world, being used not only in the US and France, but also Belgium, Sweden, Germany, Japan and elsewhere. The UK finally abandoned its gas-cooled reactor designs in the 1980s and went for PWRs. A publicly funded programme of 10 new plants was proposed, but in the end only one was built, at Sizewell on the East coast. Russia went for a slightly different approach (graphite moderated, water-cooled reactors) as did Canada (the *CANDU* heavy water moderated reactor), and some efforts were also made, for example in the US, UK, France, Japan and Russia, to develop fast neutron breeder reactors, with mixed results. Most of these fast reactor programmes have been abandoned, although Russia is still developing breeders and more may yet emerge e.g. China has a small one.

1.3 The rise and fall of nuclear power

The expansion of nuclear power in the 1970s and 1980s (see figure 1.1), meant that, in those countries that participated, it typically supplied 20%–25% of their power, although some like France went for more. However, although Generation II designs were reasonably successful, problems began to emerge. With cheaper energy sources becoming available, notably natural gas, their economics started to look poor. The near melt down of a PWR at Three Mile Island in the US in 1979 and the Chernobyl disaster in the Ukraine in 1986, with subsequent death estimates across the region due to the latter ranging up to tens of thousands, led to an increased focus on safety, which added to the cost; see box 1.1. No new plants were ordered in the US after Three Mile Island and after Chernobyl few new projects were started anywhere. The

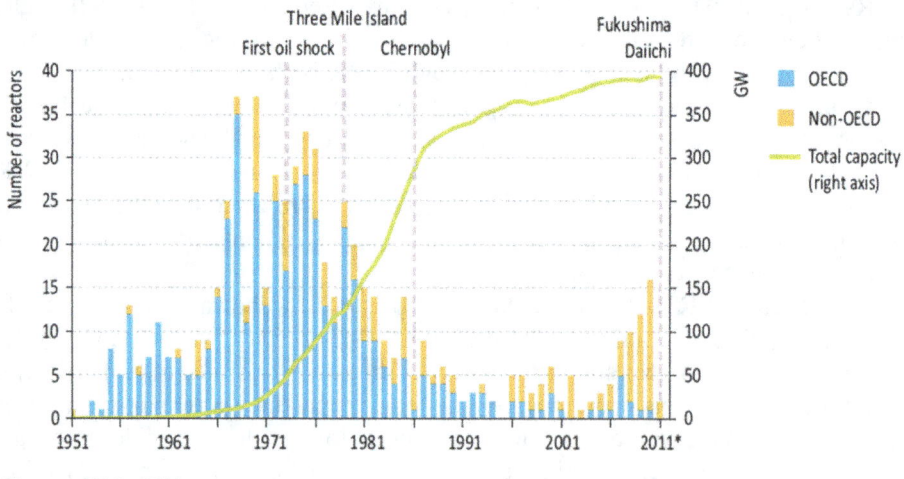

Figure 1.1. Nuclear ups and downs (OECD 2011).

boom was over, and with these two accidents being followed by Fukushima, boosting public opposition to nuclear across most of the world, and the cost of rival energy sources falling, the prospects for nuclear looked poor.

Even ignoring the accidents, the economic problems had become significant. For example, the 1200 MW PWR at Sizewell in the UK, which started up in 1995, was one of the last Generation II projects to be built. It had cost around £2.7 billion, provided by the government as a public sector project. The Conservative government was however keen to privatise the electricity sector, and sold the Sizewell plant, along with all the old AGR plants, to a newly privatised company, British Energy, for around £1.5 billion, in a heavily discounted bargain package. Unfortunately, despite this deal, British Energy ran into financial problems and had to be bailed out by the government. Eventually, in 2009, most of its assets were bought up by French company EDF. By this time, with the construction costs paid off, the plants were more economic to run, but this episode does indicate the economic problems faced by nuclear, which were worsened by the so called 'dash for gas'. The future for new nuclear plants looked very uncertain. As we shall see, that remains the case globally (BAS 2017).

An attempt at a revival had been made in the mid-2000s, partly on the basis that nuclear was low carbon (Nuttall 2005). However, it was somewhat stymied by the Fukushima disaster in 2011 (see box 1.1). Although the worst was avoided, that had major social and economic consequences in Japan. It also led some countries to abandon nuclear programmes, notably Germany (with a full phase-out set in motion), but also Italy and Switzerland, and even France decided to cut back on its nuclear reliance. With Austria, Denmark, Greece, Ireland, Norway and Portugal already anti-nuclear, within the EU, that left just the UK, and to lesser extent Finland and some Eastern Europe countries, with nuclear expansion programmes.

After Fukushima, Japan closed all its nuclear plants, although a handful have since been restarted, despite often massive local opposition. But there are no longer plans for expansion. Taiwan decided to phase its plants out. China initially halted nuclear expansion, but has since restarted it, although it is expanding renewables on a much larger scale. Investment in nuclear in China was £18 billion in 2015, dwarfed by that in renewables at over $100 billion in 2015, with renewables, including large hydro, supplying around 10 times the output of its nuclear plants and, in 2012, the output from its rapidly expanding wind power programme overtaking the output from nuclear. The situation is similar in India, with wind power output overtaking that from nuclear in 2012, while in the US, the renewables output overtook that from nuclear in 2010. Russia is the main anomaly. In terms of new projects, rather than exploiting its huge renewable resources, it has focused on nuclear, using its own technology, seeking to expand its use at home and aggressively pursuing technology export orders (WNISR 2016).

For the moment, while Russia continues along it own path, the main focus globally is on so-called Generation III designs, essentially upgraded, allegedly more 'failsafe', versions of the Generation II PWRs and BWRs, with 'passive' safety features. They include the European Pressurised-water Reactor (EPR) being promoted by the French state-owned utility EDF, the Advanced Boiling Water

Reactor (ABWR) developed by Hitachi, and the Westinghouse AP1000 developed by Toshiba. China and the US apart, the UK is one of the few countries willing to test these as yet unproven options, though their purveyors are keen to establish new markets, for example in Africa, Asia and the Middle East. So too are the purveyors of their own similar designs (notably China and South Korea), with the UK also a target (Thomas 2017).

However, the economics of these designs are far from clear, given the advent of cheaper gas and renewable generation. There have also been long delays and large cost overruns with the first two EPRs being built in France and Finland. The EPR planned for construction in the UK at Hinkley, at a cost of £24 billion, has been offered a strike price, under the UK's Contract for Difference (CfD) subsidy system, of £92.5 MWh^{-1}, which is likely to be much more than renewable projects will require by the time the EPR is running, possibly in 2025, if it goes ahead. On-shore wind and solar projects are already going ahead in the UK with CfD strike prices of around £80 MWh^{-1} and offshore wind is expected to get to around that figure soon. Indeed, a Danish offshore wind project has won a contract at around £56 MWh^{-1}.

These prices do not include grid balancing/system backup cost, which might add an extra 10%, but the costs of wind and solar technology are falling rapidly, making the current range of nuclear projects, with project costs continually rising, look increasingly uncompetitive.

In 2016, the UK government's National Audit Office noted that *'the cost competitiveness of nuclear power is weakening as wind and solar become more established. The levelised cost of electricity from wind and solar has reduced in recent years as these technologies have been deployed more widely.'* (NAO 2016). By 2025, onshore wind and large PV solar are seen as becoming significantly cheaper than new nuclear plants (BEIS 2016).

This is not just a UK or EU phenomena. Nuclear projects around the world are being abandoned as uneconomic, in the US especially (see box 1.2), with few new projects emerging, while renewables continue to boom.

Nuclear power is also unpopular with the public. After the Fukushima disaster in 2011, a global Ipsos survey put opposition at 62%, with it being much higher in some countries, e.g. 79% in Germany (IPSOS 2011). 94% voted against it in a referendum in Italy in the wake of Fukushima. Opposition around the world still remains strong. Even in countries that are still relatively pro-nuclear, opposition is quite high. For example, a 2016 Gallup poll in the US found that 54% of respondents were opposed to nuclear energy (Platts 2016). Public opinion polls can of course be wrong. There can be conflicting results, depending on what question is asked (Bisconti 2016). However, it seems very clear from all the opinion surveys that have been done that renewables are very much more popular, scoring in the 70%–80% range.

Given these social and economic trends, critics have suggested that nuclear power has little future as a response to climate change. The 2016 edition of the *World Nuclear Industry Strategy Review* commented: *'While no energy source is without its economic costs and environmental impacts, what has been seen clearly over the past decade, and particularly in the past few years, is that choosing to decarbonize with*

> **Box 1.2. Nuclear closures in the US.**
>
> Nuclear plants around the world have been closing due to their poor economics. The 2016 edition of the *World Nuclear Industry Status Report* says this has been very clear in the US (WNISR 2016).
>
> In a roundup, it notes that, in 2013, the Kewaunee plant was shut down, despite operator, Dominion, having upgraded it and, in 2011, having obtained an operating license renewal valid until 2033. Two reactors at San Onofre then followed, when replacement steam generators turned out to be faulty. Vermont Yankee shut down at the end of 2014 and early shutdown plans also hit Pilgrim and Fitzpatrick, likely to close before the end of 2017 and 2019. Exelon, the largest US nuclear operator, announced in June 2016 that it was retiring its Clinton (1065 MW) and Quad Cities (2 × 940 MW) plants in 2017 as they had been losing money for some years. PG&E in California said they would close the two Diablo Canyon units by 2025, in a slow phase out, allowing the capacity to be replaced by renewables/energy saving. In June 2016, Omaha Public Power District Board voted unanimously to shut the Fort Calhoun reactor. A board member said it was *'simply an economic decision'*. The WNISR report also quoted Nuclear Energy Institute's president as saying: *'If things don't change, we have somewhere between 10 and 20 plants at risk'*.
>
> However, there are some new plants being built and one start up so far. TVA's Watts Bar plant went online in 2016. There are also plans to keep some old plants going with extra subsidies, e.g. in New York state (WNN 2016). This subsidy approach was not seen as ideal by all (Bradford 2016, Judson 2016). It certainly seems unlikely to be sustainable long-term, as renewable costs fall further.

nuclear turns out [to be] an expensive, slow, risky and potentially hazardous pathway, and one which few countries are pursuing' (WNISR 2016).

That may be a premature obituary, but, in addition to the early retirement of many old plants in the US and elsewhere, the experience with new plants, like the EPR, has certainly not been very positive so far. Initially there were plans for seven EPRs in the US, but as markets tightened and the problems with the EPR projects in Finland and France worsened, with long delays and faults trebling the initial cost estimates, the US EPR plans were abandoned.

The situation has become even more challenging since then. What seem like possible generic faults, including in the casting of some of the steel reactor shells, led to safety checks on the new EPR castings and also retrospective tests on other earlier EDF/Areva projects, with, in late 2016, many French plants having to be taken off-line for the duration (Reitenbach 2016).

However, there are still hopes within the nuclear industry that other Generation III projects will do better in economic terms than the EPR, in the UK or elsewhere. China is clearly trying, with a large nuclear programme, aimed at doubling its nuclear contribution from 2% of electricity to around 4%, but it has had problems with some of its new Generation III projects, including not only its version of the EPR, but also its version of the AP1000 (Thomas 2016). Even so, it clearly wants to expand and develop its own technology, and so does South Korea, which has recently started up the first of its own 'APR1400' designs.

In the US, there have been delays and cost over-runs with an ABWR project and, although one new plant has started up (TVA's Watts Bar), its nuclear programme certainly does seem to be stalling (Kee 2016). In the UK, while the old plants are being closed, the proposed EPR at Hinkley is still going ahead, with some Chinese funding, along possibly with another EPR at Sizewell. There are also proposals for Japanese ABWRs at Wylfa and Oldbury and an AP1000 at Moorside, next door to the Sellafield/Windscale complex, despite strong local opposition and adverse independent technical commentary (Roche 2016). There is also a proposal to build a Chinese-designed reactor at Bradwell in Essex. That too is being strongly opposed locally (Blowers 2016). Given the involvement of China, UK national security concerns have been raised. But if it can get UK regulatory clearance, a *Hualong One* design plant may be built, the first outside of China. The government's overall aim is for nuclear to generate around 30% of UK power by 2035 (Neville-Rolfe 2016).

Some see the UK nuclear programme as misguided, not least since, as a contributor to a House of Lords review of nuclear economics put it: *'We will be building four different reactor types, with at least five different manufacturers, simultaneously. That is industrial insanity.'* While it might also be seen as a brave attempt to test the waters with a range of designs, that could prove very expensive. That does seem to be one lesson from the past (Carter 2016).

1.4 On to Generation IV?

It remains to be seen what will become of these and other Generation III projects. However, given their limitations and problems, there have been calls for a more radical approach, and the development of new, hopefully more cost effective, nuclear technology. A range of so-called Generation IV options has been proposed to underpin the hoped for Nuclear Renaissance. Some of them are in fact old ideas, now being re-explored. Indeed, it could be argued that, with the industry apparently at something of a dead end with current types of reactor, it has had to go back to see if anything can be rescued from the past.

Although, put more positively, it might be seen as taking a fresh look at what might be possible, it does seem to drawing heavily on what went before. For example, the so-called 'pebble bed' helium gas cooled high-temperature reactor has received some attention. The fuel is encapsulated in graphite balls, sitting in a hopper, rather than in rods in channels. This feature allows spent fuel to be removed more easily—old pebbles are simply extracted. There were plans for one to be developed in South Africa. However, after £800 million in expenditure, the project was abandoned in 2010 due to funding problems. China then took the concept over, with a 200 MW prototype planned, but not yet running.

It is not a new idea. High temperature helium-cooled reactors were tested in the 1960s/1970s in the UK (the 20 MW Dragon) Germany (the 15 MW AVR) and the US (Peach Bottom, Fort St Vrain), but were not followed up, in part due to operational/safety issues, which some said would also apply to the pebble bed design (Thomas 2008). Certainly high temperature reactors have had a checkered history, with none so far having been commercially viable and all having been shut down before the end of their planned life (Ramana 2016).

As we shall see, many other early ideas are now back on the agenda. Certainly, the early burst of innovation in the 1950s and 1960s, had thrown up many different ideas which some say might prove to be worth re-exploring, rather than simply trying to soldier on with upgraded Generation III versions of Generation II designs. The earlier phase had been typified by optimism and experiment, somewhat different from now. However, belief in the transformative power of technology generally is still strong, having been enhanced by the contemporary success of information, communication and computer based technologies, though, it is wise to avoid hubris and looking back may add some perspective.

While it is perhaps natural to have positive expectations of new technology, in the nuclear field this has not proven to be justified in some cases. In 1965, Fred Lee, the UK's then Minister of Power, had famously told the House of Commons: *'We have hit the jackpot this time'*, with the AGR. Unfortunately, deployment went very wrong as the AGR programme unfolded. The first station, on the south Kent coast, was Dungeness B. It was ordered in August 1965, but did not start up until December 1982, over 17 years later, by which time its cost had reached more than five times the 1965 estimate, and its intended output had been scaled down more than 20%. In 1985, two decades after the original order, the second reactor at the station had only just started up. Atomic Power Constructions, the company that won the Dungeness B contract in 1965, had by 1970 collapsed in total disarray, technical, managerial and financial (Patterson 1985).

Project disasters like that might be seen as part of the learning process, and, as we shall see, there certainly have been many problems in the past, some of them quite dramatic. However, technical problems, and even major setbacks, are not uncommon in the innovation field, and reactor development programmes continued around the world, with their problems and progress issues, as with the more recent disasters mentioned above, offering warnings and insights, for example on safety concerns. Although some take time to emerge (Thomas 2012).

The next section looks back at the early days to see what can be learnt from what happened, how things were done then and if any of the ideas are still relevant. The institutional context for innovation is different now, with 'high technology' projects often requiring major finance and large teams, centrally directed. Certainly some still hanker back to the time when allegedly it was all much less bureaucratic. President Eisenhower, in his famed 1961 warnings about the rise of the 'military–industrial complex', also bewailed the fact that *'today, the solitary inventor, tinkering in his shop, has been overshadowed by task forces of scientists in laboratories and testing fields'*. That may be overly romantic, but he may have been right to note that *'partly because of the huge costs involved, a government contract becomes virtually a substitute for intellectual curiosity'* (Eisenhower 1961).

Whether that situation can be changed remains to be seen. The nature of nuclear technology seems to make it hard, much harder than for the relatively less complex renewables. For example, modern wind turbines evolved very successfully in Denmark in the 1970s initially from simple designs developed on a local agricultural engineering craft basis (Karnoe 2006). It was similar for solar heat collectors and biogas digesters. Although novel wave and tidal stream projects still seem to be the province of small start up companies in the UK and elsewhere, the DIY, backyard, small team approach may now be over for most other renewables and it certainly seems to be over for nuclear. However, some nuclear enthusiasts evidently think that it might possible, and indeed necessary, to recapture the more open-ended creative ethos that it seems once existed. Some see it as a Golden Era. Certainly the incentive was there to be inventive and come up with radically new technology. As we shall see, there were many possible lines of development, each attracting their own proponents. In 1945 Enrico Fermi said: *'The country which first develops a breeder reactor will have a great competitive advantage in atomic energy'*, with plutonium production emerging as a front runner, but initially there was all to play for, and apparently, few limits on imagination.

References

BAS 2017 *Global Nuclear Power Database* Bulletin of the Atomic Scientists interactive chart http://thebulletin.org/global-nuclear-power-database

BEIS 2016 *Electricity Generation Costs* Department for Business, Energy & Industrial Strategy, London https://www.gov.uk/government/uploads/system/uploads/attachment_data/file/566567/BEIS_Electricity_Generation_Cost_Report.pdf

Bisconti A 2016 *Public opinion on nuclear energy: what influences it* Bulletin of the Atomic Scientists, April 27th http://thebulletin.org/public-opinion-nuclear-energy-what-influences-it9379

Blowers A 2016 *Bradwell site unsuitable and unsustainable*, No 2 Nuclear Power web post, October 4th http://www.no2nuclearpower.org.uk/recent-additions/bradwell-site-unsuitable-unsustainable

Bradford P 2016 *Compete or suckle: Should troubled nuclear reactors be subsidized?* The Conversation, August 18th https://theconversation.com/compete-or-suckle-should-troubled-nuclear-reactors-be-subsidized-62069

Carter C 2016 *Nuclear industry in the UK: Back to the future?* Web Blog, Science Policy Research Unit, University of Sussex, September 28th http://blogs.sussex.ac.uk/policy-engagement/2016/09/28/nuclear-industry-in-the-uk-back-to-the-future/

Drogan M 2016 *The Nuclear Imperative: Atoms for Peace and the Development of U.S. Policy on Exporting Nuclear Power, 1953–1955* Diplomatic History **40** Issue 5 948–974 http://dh.oxfordjournals.org/content/40/5/948.abstract?sid=ebac56d5-f3a7-45e0-a586-f6f0294a6930

Dwyer P 2007 *Windscale: A nuclear disaster* BBC News report http://news.bbc.co.uk/1/hi/sci/tech/7030281.stm

Eisenhower D 1953 Press Release, *Atoms for Peace* Speech, Dec. 8th, Dwight D Eisenhower Presidential Library http://www.eisenhower.archives.gov/research/online_documents/atoms_for_peace/Binder13.pdf

Eisenhower D 1961 *Military-Industrial Complex Speech* public policy papers archive http://coursesa.matrix.msu.edu/~hst306/documents/indust.html

Elliott D 2013 *Fukushima: Impacts and implications* (Basingstoke: Palgrave) http://www.palgrave.com/page/detail/fukushima-david-elliott/?K=9781137274328

Fairlie I 2014 *New UNSCEAR Report on Fukushima: Collective Doses* Dr Ian Fairlie website, April 2nd http://www.ianfairlie.org/news/new-unscear-report-on-fukushima-collective-doses/

Fairlie I and Sumners D 2006 TORCH: *The Other report on Chernobyl* produced for The Greens/EFA in the European Parliament http://www.chernobylreport.org

Guardian 2011 *Nuclear power plant accidents: listed and ranked since 1952*, Guardian Data Blog https://www.theguardian.com/news/datablog/2011/mar/14/nuclear-power-plant-accidents-list-rank#data

IAEA 2006 *Chernobyl's Legacy: Health, Environmental and Socio-Economic Impacts and Recommendations to the Governments of Belarus* The Russian Federation and Ukraine, The Chernobyl Forum: 2003–2005, International Atomic Energy Agency, Vienna http://www.unscear.org/unscear/en/chernobyl.html

IPSOS 2011 *Global Citizen Reaction to the Fukushima Nuclear Plant Disaster*, IPSOS Global Advisor, global poll carried out in May, published in June http://www.ipsos-mori.com/Assets/Docs/Polls/ipsos-global-advisor-nuclear-power-june-2011.pdf

ISRN 2011 Radiation fallout plot over Europe during Chernobyl incident (1986) animated map of Cesium 137 levels https://www.youtube.com/watch?v=_USpAPkAd5A

Judson T 2016 *Too Big to Bail Out* Nuclear Information and Resources Service, Tokama Park, MD, November https://drive.google.com/file/d/0B2b-a19f3q6ZdnJuajE1ckxKa2c/view

Karnoe P 2006 *Technological innovation and industrial organization in the Danish wind industry*, Entrepreneurship & Regional Development **2** 105–124 http://www.tandfonline.com/doi/abs/10.1080/08985629000000008

Kee E 2016 *US nuclear industry in decline* Nuclear Engineering International, January 28th http://www.neimagazine.com/features/featureus-nuclear-industry-in-decline-4498254/

Morelle R 2007 *Windscale fallout underestimated*, BBC News report http://news.bbc.co.uk/1/hi/sci/tech/7030536.stm

NAO 2016 *Nuclear Power in the UK*, National Audit Office, London http://www.nao.org.uk/report/nuclear-power-in-the-uk

Neville-Rolfe L 2016 *UK statement to the IAEA international conference on nuclear security*, Energy Minister Baroness Neville-Rolfe https://www.gov.uk/government/speeches/uk-statement-to-the-iaea-international-conference-on-nuclear-security

Nuttall W 2005 *Nuclear Renaissance: Technologies and Policies for the Future of Nuclear Power* (Bristol: IOP Publishing)

OECD 2011 World Energy Outlook, OECD, Paris https://www.iea.org/publications/freepublications/publication/WEO2011_WEB.pdf

Patterson W 1985 *Going Critical: An Unofficial History of British Nuclear Power* (London: Paladin Books/Friends of the Earth) http://www.foe.co.uk/sites/default/files/downloads/going_critical.pdf

Pease R 2008 *The Story of Britain's Sputnik*, BBC News article, January 15th http://news.bbc.co.uk/1/hi/sci/tech/7190813.stm

Platts 2016 *Majority in US opposed to nuclear power: Gallup poll*, Platts News, Washington DC, March 18th http://www.platts.com/latest-news/electric-power/washington/majority-in-us-opposed-to-nuclear-power-gallup-21124528

Ramana M V 2016 *The checkered operational history of high-temperature gas-cooled reactors*, Bulletin of the Atomic Scientists http://dx.doi.org/10.1080/00963402.2016.1170395

Reitenbach G 2016 *France's Nuclear Storm: Many Power Plants Down Due to Quality Concerns*, Power magazine, November 1st http://www.powermag.com/frances-nuclear-storm-many-power-plants-down-due-to-quality-concerns/

REN21 2016 *Renewables 2016 Global Status Report*, Renewable Energy Network for the 21st century http://www.ren21.net/status-of-renewables/global-status-report/

Roche P 2016 *The AP1000 Nuclear Reactor Design*, Edinburgh Energy and Environment Consultants report, November http://www.no2nuclearpower.org.uk/wp/wp-content/uploads/2016/11/AP1000_reactors.pdf

Ten Hove J and Jacobson M 2012 Worldwide health effects of the Fukushima Daiichi nuclear accident *Energy Environ. Sci.* **5** 8743–8757 http://pubs.rsc.org/en/content/articlelanding/2012/ee/c2ee22019a

Thomas R 2012 *The Nuclear Disaster of Kyshtym 1957 and the Politics of the Cold War*. Environment and Society Portal, *Arcadia*, No. 20. Rachel Carson Center for Environment and Society http://www.environmentandsociety.org/node/4967

Thomas S 2008 *Safety issues with the South African Pebble Bed Modular Reactor: When were the issues apparent? A briefing paper*, Public Services International Research Unit, University of Greenwich, July http://www.mnet.co.za/Mnet/Shows/carteblanche/story.asp?Id=3516

Thomas S 2016 *China's nuclear roll-out facing delays*, China Dialog, October 26th http://www.chinadialogue.net/article/show/single/en/9341-China-s-nuclear-roll-out-facing-delay

Thomas S 2017 China's nuclear export drive: Trojan Horse or Marshall Plan? *Energy Policy* **101** 683–691 http://www.sciencedirect.com/science/article/pii/S0301421516305031

WNA 2017 *World Nuclear Power Reactors & Uranium Requirements*, World Nuclear Association, London http://www.world-nuclear.org/information-library/facts-and-figures/world-nuclear-power-reactors-and-uranium-requireme.aspx

WNISR 2016 World Nuclear Industry Status Report, annual independent survey and review, London http://www.worldnuclearreport.org/

WNN 2016 *New York State sets out subsidy proposals*, World Nuclear News, July 12th http://world-nuclear-news.org/EE-New-York-State-sets-out-subsidy-proposals-1207167.html

IOP Concise Physics

Nuclear Power
Past, present and future
David Elliott

Chapter 2

Nuclear innovation: the early days

The early days of nuclear power development were typified by a burst of optimistic, open-ended experimental exploration of options, with a wide range of designs and types of reactor system being tested in the USA and elsewhere. This chapter explores the early days and their implications for the present.

2.1 Early US experiments

In the summer of 1956, a handful of men gathered in a former little red schoolhouse in San Diego. They were some of the most imaginative scientists and engineers of their generation. They included ex-Manhattan Project physicists like Edward Teller, and Freeman Dyson, from the Institute for Advanced Study in Princeton.

According to a retrospective account in *Scientific American* in 2013, the team '*benefited from a maximum of free inquiry and individual creativity and a minimum of bureaucratic interference. There was no overarching managerial body dictating the thoughts of the designers. Everyone was free to come up with any idea they thought of, and the job of the rest of the group was to either refine the idea and make it more rigorous and practical or discard it and move on to the next idea*' (Jogalekar 2013). And they came up with a viable small reactor design—TRIGA, this standing for 'Training, Research, Isotopes, General Atomics'.

It seems 70 of them were built for use primarily to produce isotopes for scientific and engineering experiments. It was a successful and evidently lucrative niche application, but the *Scientific American* article lamented the fact that this creative process seems to be less apparent now, just when radical innovation is needed to get out of the impasses nuclear energy development has met.

It quoted Dyson, who in his memoir *Disturbing the Universe* says: '*The fundamental problem of the nuclear industry is not reactor safety, not waste disposal, not the dangers of nuclear proliferation, real though all these problems are. The fundamental problem of the industry is that nobody any longer has any fun building reactors… Sometime between 1960 and 1970 the fun went out of the business. The adventurers, the experimenters, the inventors, were driven out, and the accountants and managers took control. The accountants and managers decided that it was not cost effective to let bright people play with weird reactors. So the weird reactors disappeared and with them the chance of any radical improvement beyond our existing systems. We are left with a very small number of reactor types, each of them frozen into a huge bureaucratic organization, each of them in various ways technically unsatisfactory, each of them less safe than many possible alternative designs which have been discarded. Nobody builds reactors for fun anymore. The spirit of the little red schoolhouse is dead'* (Dyson 1981).

The *Scientific American* article did however go on to point to a few new projects which it saw as similarly creative, including the travelling wave reactor design being developed by private company Terrapower. Certainly there are now several new US startup companies exploring ideas for small modular reactors, molten salt reactors, the use of thorium fuel and fast neutron systems.

Many of these ideas involve going back over old ground. For example, in the early years, the US did try to develop small reactors, initially for the military and also for civilian use, with mixed success. From 1946 to 1961, the US Air Force spent over $1 billion trying to build small compact reactors, including a novel uranium-based 2.5 MW molten salt reactor, to power long-range bombers, before abandoning the idea. The Navy was more successful with power plants for aircraft carriers and submarines. However, as Ramana notes in a review of early US small modular reactors (SMRs): *'These have quite different requirements from today's SMR proposals. A submarine reactor is designed to operate under stressful conditions - to provide a burst of power when the vessel is accelerating, for example. And unlike civilian power plants, naval nuclear reactors don't have to compete economically with other sources of power production. Their overwhelming advantage is that they enable a submarine to remain at sea for long periods of time without refueling'* (Ramana 2015).

The US Army programme was more relevant. As Ramana describes, it led to the construction of eight small reactors, some of them tested in remote sites, although there were problems. For example, the PM-3A at McMurdo Station in Antarctica, developed cracks and leaks, leading to significant local contamination. In 1976 the Army cancelled the programme. It was not seen worth the expense of pursuing it, since diesel power packs were available and cheap.

The US Atomic Energy Commission (AEC) was however keener to persevere with small test plants, and 17 reactors with power outputs of under 300 MW were commissioned. There were many test runs, with mixed results, but the main aim of the AEC programme was to develop prototypes for subsequent expansion to larger projects for civil power use. The AEC supported three projects like this in its 1955 programme. Of these, the 61 MW Fermi 1 prototype fast-breeder neutron reactor, which was near Detroit, is the best known, since it suffered a partial core meltdown in 1966 (Fuller 1975). Ramana says that the other two reactors were relatively successful in meeting their goals. The 185 MW Yankee operated for 31 years, but its decommissioning took 16 years and cost $608 million.

As a follow up, in 1955, the AEC announced a second round of funding, but this time with small reactors as the goal, not an initial step. The programme included a

22 MW reactor in Elk River, near Minneapolis, and the La Crosse boiling water reactor in Genoa, Wisconsin. Elk River was a variant of the boiling water reactor, but used a mixture of highly enriched uranium and thorium. Construction began in 1959, but it was not declared as operating commercially until 1964, due to various engineering problems, including cracks in some components. They evidently contributed to its remarkably short operating life: just three and a half years. The decommissioning process took three years and cost $6.15 million, which was almost the same figure as the initial estimate for construction. The 50 MW La Crosse boiling water reactor was more successful, operating for 18 years, but evidently at high cost.

Since then, Ramana reports, no more small reactors have been commissioned in the US, with the exception of the 330 MW Fort St Vrain Nuclear Generating Station in Platteville, Colorado, mentioned earlier. It was an experimental high-temperature gas-cooled reactor design deemed to be 'ultrasafe'. It started up in 1976, but rarely operated at full capacity only delivering around 15% of the theoretical output, and it was shut down in 1989.

Now however small modular reactors of various kinds are back on the agenda, in part in the belief that they may be better suited to the more flexible power systems that are emerging given the advent of decentralised energy generation using renewables and local energy sources. There is also interest in thorium and fast neutron reactors, possibly as small modular units. The next section looks at what was attempted in the early days, at various scales, in the USA and elsewhere, in relation to thorium and fast neutron systems, moving on to the present situation.

2.2 Thorium reactors and fast breeders

As noted above, a thorium fuel mix was used in one early US plant and the Fermi plant was a prototype commercial liquid sodium fast breeder, following on from a series of experimental projects. The US's first commercial reactor at Shippingport, which started up in 1957, also, later on, used some thorium. In addition to the 1950s Airforce project mentioned above, the US also did some work on the use of molten salt fuel and thorium in the 1960s.

This was at Oak Ridge National Laboratory (ONL), where an experimental 7.4 MW molten salt reactor using uranium was designed to pave the way for a system using thorium. It ran for four years, but, as noted in box 2.1, this line of work was discontinued in favour of work on fast-breeders, some say because it did not lead to the production of plutonium for weapons. Others, in contradiction, say it was because the use of thorium did involve a potential (if tricky to handle) weapons material: uranium 233. Whatever the actual reason, molten salt and thorium work as promoted by ONL and nuclear pioneer Alvin Weinberg, gave way to breeders, which, for a while, were heavily promoted in the US (Cochran *et al* 2010).

Enthusiasm for fast breeders did not last, with proliferation issues beginning to be a concern. The US breeder programme had included an experimental EBR-1 and then a 20 MW sodium-cooled EBR-II reactor, which started operating at Idaho National Laboratory in 1964. In the 1980s it was used to test the Advanced Liquid

> **Box 2.1. Early US molten fluoride salt reactor work.**
>
> The *Energy from Thorium* website reports as follows: '*A small, proof-of-principle liquid-fluoride reactor was built and operated in 1954 at Oak Ridge, and two years later under the encouragement of laboratory director Alvin Weinberg, a more significant examination began of liquid-fluoride reactors for electrical generation at terrestrial power stations. Weinberg also encouraged the examination of the thorium fuel cycle implemented in liquid fluoride reactors, and this work led to the construction and operation of the Molten-Salt Reactor Experiment (MSRE) at Oak Ridge. The MSRE operated from 1965 to 1969, when it was shut down under the orders of Milton Shaw of the Atomic Energy Commission so as to free up additional funding for the liquid-metal fast breeder reactor (LMFBR) program. The molten-salt program continued for another three years at Oak Ridge until it was cancelled in 1972 under Shaw's orders*' (Energy from Thorium 2016).
>
> According to another report, in 1972 Oak Ridge had in fact proposed a major new development programme, the Molten Salt Breeder Experiment, with a total programme cost estimated at $350 million over 11 years. That was turned down, as was a follow up bid in 1974, and, in 1976, ORNL was ordered to finally shut down the MSBR programme (Cochran *et al* 2010). A 1969 Oak Ridge National Lab film looks at the earlier project work (Thorium Energy World 2016a).
>
> The Weinberg Foundation provides an account of some of the problems that were faced with the proposed use of thorium in molten salt reactors: '*As thorium is chemically similar to rare-earth fission products, it is hard to process out those fission products from the fuel salt without also removing the thorium. The early 1960s saw the development at ORNL of two-fluid designs which sought to overcome this problem by keeping the thorium separate from the fuel salt. To achieve this, the two-fluid designs employed a graphite barrier between the core and blanket salts. The separation of fluids necessitated complex plumbing design involving multiple graphite tubes. During the life of the ORNL molten salt reactor programme the plumbing design problem was never resolved.*'
>
> Of course it did not have to be, as no thorium was used at this stage, just uranium and (the resultant) plutonium. However, later on, for the proposed, but in the event not followed up, Molten Salt Breeder programme, the idea was to go for a 'single fluid' approach since new extraction techniques had emerged (Weinberg 2013). But as we shall see later (box 3.2), a 'two fluid' fissile core and fertile thorium blanket approach has been proposed for the contemporary Molten Fluoride Thorium reactor.

Metal Reactor/Integral Fast Reactor concept (with integral fuel reprocessing), but that work was discontinued in 1994, three years before completion. The whole breeder programme was wound up in 1997 by President Clinton. It left behind an unedifying history, including the 12-year Clinch River Breeder Reactor saga, Once touted as the way ahead, after escalating costs (to $8 billion), it was eventually cancelled.

It was a similar story around the world; see box 2.2. Although Russia, India and Japan soldiered on, elsewhere fast breeder work ground to a halt in the 1990s, given technical and economic problems, and concerns about weapons proliferation. There

Box 2.2. Sodium-cooled fast reactors: early days and costing estimates.

Sodium-cooled fast reactors have had a mixed history. Some have operated well, while others have done poorly, with the partial meltdown of the Fermi plant near Detroit in 1966, mentioned earlier, being perhaps the most dramatic example. In addition to the directly nuclear-related aspects of fast neutron reactors (one problem being that fission is harder to control), working with liquid sodium can be risky. Sodium catches light in air and reacts strongly with water, generating hydrogen gas which can explode. Sodium leaks and fires have bedeviled most fast reactor projects. Japan's Monju reactor, commissioned in 1994, and connected to the grid in 1995, had a sodium leak and fire in 1995. It was closed until May 2010, when it was restarted for testing, but suffered another accident in August 2010. It has not been restarted since, and in 2016 it was decided that it will be decommissioned.

The UK started work on fast breeders at Dounreay in Scotland in 1955, but abandoned it in 1994, in part since, with nuclear power generally in decline at that point, shortage of uranium, one of the justifications for breeders, was no longer an urgent issue. There had also been many delays and reliability problems with its second 240 MW plant, and post-closure site clean-up has proved hard.

In the US, Experimental Breeder Reactor I started up in 1951 and, also in Idaho, the larger EBRII in 1964, this later being used to test the integral fast reactor concept. However, amidst some dissention, the US abandoned fast breeder work in 1997, with concerns about plutonium proliferation being cited.

France had a major fast breeder programme up until 1998. Its Phenix plant had allegedly been used to make its bomb. The follow-up Superphenix plant had many operational problems, and reportedly, only operated with an average capacity factor of under 7% over the 11 years before it was shut.

Russia started work on fast breeders in 1955, and despite some sodium fires and long delays, has continued its work on them. It is currently developing models using molten lead as coolant. India was a somewhat later entrant in the field, but despite large ambitions, the programme has had many problems, including sodium fires and long delays (Ramana 2016).

The economics of fast reactors continues to be an issue. So far, up to $100 billion has been spent on them around the world, and although most have been prototypes, with as yet none having proved commercially viable, we do have some idea of their costs per installed kilowatt. They seem to have increased over time, ranging from around $4000/kW for the US Fermi plant up to £20 000/kW for Superphenix (in 1996 $), with the capital costs/kW being seen as typically over twice that of water-cooled reactors of similar capacity. Though mass production could reduce that, it has been suggested that breeders were still likely to cost 25% more than water-cooled plants (Cochran *et al* 2010).

was strong public opposition, for example in Germany: the idea of producing more plutonium was not welcomed.

The engineering, with small cores cooled with molten sodium, was also difficult. Looking back over the experience, Ramana says: *'The lesson from the many decades of such pursuit has been that these reactors are expensive, are prone to operational problems and sodium leaks, and are susceptible to severe accidents under some circumstances'* (Ramana 2013).

The early days of reactor development in the US certainly had dramas, with fast reactor problems, as well as problems with other projects. In addition to the already mentioned partial fuel meltdown at the Fermi fast neutron plant, and a meltdown at EBR-1 in 1955, there were major problems with some other projects, for example with the Simi Valley Sodium Reactor in 1959, the subject of a *History Channel documentary* (Sim Valley Disaster 2014) and the 3 MW experimental SL-1 reactor at the US National Reactor Testing Site in Idaho, which exploded in January 1961, with the fuel melting down, killing three operators. A grim forensic video was produced (AEC 1961).

Tragic episodes like these, and similar ones elsewhere (e.g. at Chalk River in Canada, where there was a partial fuel meltdown, hydrogen explosion and site contamination in 1952 and a fuel fire and site contamination in 1958), did not slow programmes in the US or elsewhere. Although, as noted above, fast breeders were mostly sidelined later on, a wide range of projects emerged over the years, including high temperature reactors and the use of thorium as a fuel in conventional plants. For example, starting in 1967 Germany ran the 15 MW Atom Versuchs Reacktor (AVR) using a thorium-based fuel for 95% of the time and Canada ran tests in three research reactors and one pre-commercial reactor with fuels combining uranium and thorium. In India, the Kamini 30 kW experimental neutron-source research reactor, which uses U233 as fuel, was started in 1996. It was built next to a 40 MW fast breeder test reactor, in which thorium was irradiated to make the U233 (Global Data 2011).

As we shall see, thorium based systems are now quite high on the agenda and the fast reactor concept, including the integral fast reactor (IFR), with integral fuel reprocessing, is also popular again in some circles, especially if it can run on its own bred fuel and be used to burn up fission by-products, in effect running on what are now considered to be wastes (Till and Chang 2011). Current IFR proponents include Australian academic Professor Barry Brook (Brook 2016). A similar line has been taken forcefully over the years by nuclear advocate Kirk Sorensen: he says waste is actually not waste but very valuable; see his 2010 Google Tech Talk presentation (Sorensen 2010). He is also a strong advocate of thorium molten salt reactors, which he sees as also being able to burn waste (Sorensen 2009).

Adopting a similarly bullish approach, the Weinberg Foundation says the molten salt reactor is a *'revolutionary'* advance for fission: it is *'extremely fuel efficient, generates very little waste, and offers unique passive safety features. Crucially, the MSR has outstanding load-following capability and will provide a low-carbon alternative to gas as a flexible source of electricity to support renewables'* (Weinberg 2014).

With Microsoft's Bill Gates also being keen on novel nuclear technology and asking what would happen if *'instead of burning a part of uranium, the one percent, which is the U235, we decided, let's burn the 99%, the U238'* (Gates 2010), we may yet see a return to these ideas, alongside, and maybe as part of, the recent revival of interest in small reactors.

The nuclear industry clearly needs a shot in the arm if it is to survive much less expand. Not everyone will think that it should be revived, or that these new technology ideas will be enough. But efforts are certainly being made to get new technology developed. Box 2.3 briefly reviews the existing and new options and some of their pros and cons, before we move on to look at what is actually being done to develop the new technology options.

Box 2.3. Reactors types—a *simple* summary of pros and cons.

Generation 1 The early days of full scale plant operation
Some early plants were graphite moderated, and the Windscale plutonium production piles were air cooled. Blowing air over a stack of hot metal resting in what are in effect charcoal blocks was perhaps not a good idea, as proved to be the case when one pile overheated, and, BBQ-like, caught fire. Switching to the use of inert carbon dioxide gas for cooling, as in the UK's subsequent Magnox and AGR reactors, was more sensible. In the Russian RBMK design, as used at Chernobyl and elsewhere, each fuel rod was separately contained in its own tubular pipe—an amazing feat of plumbing. In theory that made it less prone to catastrophic loss of coolant and containment accidents. But not, it seems, to the deliberate disconnection, for a test at Chernobyl plant 4, of all the safety interlocks, to see what would happen if there was an external power cut. It was expected that the plant, run at low power, could produce enough electricity to keep running safely. Instead, it ran out of control and exploded.

Generation II The commercial expansion period
Water-cooled reactors, like the PWR and BWR, have a big advantage over gas-cooled systems in that water is denser and has much higher heat retention capacity. If pressurised, as in PWRs, the boiling point is raised, so that higher temperatures can be reached, increasing heat transfer capacity further. To do this, a cooling loop is used, transferring heat from the pressurised reactor core to a heat exchanger in a low pressure boiler outside, with water then being boiled for power production. However, if pressurisation is suddenly lost, e.g. through a rupture of containment, the coolant flashes off into steam, cooling is lost and the reactor overheats and can melt down. In BWRs, the water boils at more normal temperatures, actually in the reactor core. That is a little worrying: it may be activated. But since there is no secondary cooling loop or heat exchanger, it is cheaper. However, since BWRs run at lower coolant temperature, they need more cooling water throughput. When power for running the cooling pumps for the BWRs at Fukushima was lost, vast amounts of sea water had to be pumped through, using fire tenders and the like, to stop further meltdowns. Not a brilliant idea: it is corrosive and became radioactive. Some was dumped in the sea, some has been kept in a large number of tanks on site. Lesson: do not put the back-up diesel power units below flood level. Better still: do not build nuclear plants in earthquake zones.

Generation III The current revisions
The safety problems revealed by earlier models led to the addition of ever more complex and expensive interlocks and controls, plus more backup emergency cooling systems. In Generation III reactors an attempt is made to seek to reduce some the costs and unreliability of that by redesigning operations so that they are 'fail safe' with passive safety features, e.g. instead of extra cooling pumps, large tanks of water are placed above the reactor to flood it by gravity should power for pumps fail. As a last

ditch (almost literally) some have 'core catchers' built into the basement; reinforced traps designed to stop a melted core from burning through the concrete floor into the top soil underneath. Few Generation III plants are running as yet, so it is not clear if the new features will work well.

Generation IV Candidate new reactors
Apart from the experiences with various early prototypes discussed in the main text above, we have even less knowledge about how these might fare, but there are some basic nuclear physics issues which may help in assessing the viability of the various approaches that may be taken. A key point concerns the relative effectiveness of fast and slow neutrons in causing *fissions* in U235/U233/Pu239, as against being *absorbed* by U238 and other atoms. If we want the maximum number of fissions, we go for slow moderated neutrons, as in so-called thermal reactors, including those using thorium (or rather U233) as a fuel. If we want to breed more plutonium, we go for fast neutrons, as in IFRs, with Pu239 fission also yielding more neutrons than U235 fission. However, while fast or slow neutrons can also convert thorium to U233, in the process, an interim isotope, U234, is also produced by neutron absorption. It is not fissile. Ideally we want to avoid that, to maximise U233 production, and experts are apparently divided on whether that can be done effectively. The use of molten salts may help with some of these problems, perhaps making it easier to play with the nuclear chemistry and tap off unwanted by-products, but it is far from proven. Some current trends are looked at below.

Generation V? The far future—fusion
Apart from indirect understanding of what happens in the Sun and other stars (and H bombs) we have no experience of fusion, but we do know that, depending on how it is done on Earth (or in space), it will generate neutrons and gamma radiation, as well as some radioactive tritium. There may also be operational issues. In some variants, fusion can only be achieved in pulses, not continually.

The economic problems facing Generation III are an obvious driver for change, but there are others, most notably safety and also fuel efficiency. Uranium reserves are finite. There is sufficient uranium available to run conventional thermal reactors for many decades, but if the use of nuclear energy is to expand, using conventional reactors, the high-grade resource will dwindle, some say dramatically (Energy Watch 2006). While others say potential shortfalls are a long way off (OECD-NEA 2016), they are one reason why there has been interest in thorium, and also in fast breeders, given that the latter can convert the otherwise wasted U238 into a new fuel and thus extend the lifetime of the resource significantly. If they can also burn up other wastes, that would be another key attraction, waste management being a key issue.

Not all nuclear enthusiasts think that radical new technology is the answer to the problems facing nuclear power, at least not in the short to medium term. They may be right. In the short term, some nuclear operators are looking to extend the life of existing Generation II plants, since, having paid off their construction debts, some, possibly with upgrades, may be able run competitively. However, clearly not all can, as witness the many recent closures and the call for government or consumer subsidies to keep some running. In the medium term, the fate of nuclear power may well be decided on the success or otherwise of new Generation III ideas. Many nuclear enthusiasts are hopeful that, once the initial 'first of a kind' problems have been overcome, the operational economics will prove to be good, in which case, some say, diverting resources to new expensive and possibly unproductive research programmes may be unwise. A similar view can sometimes shape reactions from fission enthusiasts faced with what they may see as overly extravagant funding for fusion. It seems the old resists the new.

Of course, as we have seen, that has not always been the case, and that formulation of the reactions to change also assumes that there is a fixed direction of travel—progress along coherent predictable lines. That is not obvious. There can be new paths and new options, not necessarily nuclear-related. Certainly change is likely. Generation III will have to be replaced by something at some point, and for many nuclear enthusiasts it has to be with a new generation of nuclear technology, Generation IV or, at the very least, upgrades of Generation III.

How soon that might happen is unclear. Most Generation IV projects are still at the R&D stage, with commercial-scale systems at best decades away. Technological breakthroughs are always possible, and some enthusiasts see rapid development and wide deployment as being urgent, but it is far from clear, at this stage, which of the options (if any) should be promoted. Being positive, perhaps the ethos apparent in the early days of nuclear (and renewable energy) development, of 'letting a thousand flowers bloom', has merit, particularly at the early stage of innovation, when costs are relatively low. Later on though choices will have to be made.

2.3 Generation IV design parameters and constraints

As box 2.3 illustrates, leaving aside fusion and the far future, there are some basic issues related to the underlying physics that may shape the choice and design of the near-term Generation IV options. They will also have to face up to the problems that confronted earlier designs, such as their relatively poor economics and high waste production.

There can be interactions between these two factors (and of course others, such as safety and security) that can limit what can be done. For example, some Generation III plants sought to improve their overall economic performance by adopting a high fuel burn-up approach. The use of more highly enriched fuel allows for the fuel rods to be left in the core longer, with more energy being obtained per fuel cycle. That improves the plants operational economics. However, it also means that what is left over at the end is much more radioactive since more fissions have occurred. So the spent fuel/waste management problems are much harder to deal with, requiring thicker shielding and longer-term storage periods before activity levels fall. That can undermine their overall economics (Edwards 2008).

Some Generation II and III designs adopted essentially the opposite approach, the use of non-enriched fuel, as in the Canadian CANDU reactor. That has the advantage of avoiding the need for expensive and difficult uranium enrichment. It uses heavy water, an isotope of ordinary 'light' water, with heavier deuterium atoms, rather than hydrogen atoms, making it better at moderating neutron speeds. So a reasonable number of fissions occur, even in lower grade U235 fuel, and, in theory, there is also a lower-activity spent fuel output that can be relatively easily processed for reuse as a new fuel. The UK tested a somewhat similar idea in the 1970s, the so-called Steam Generating Heavy Water Reactor, but it was abandoned as uneconomic compared with PWRs, and also given the problem shared with CANDU that radioactive tritium can be produced from neutron interactions with the deuterium moderator.

A revised version of the CANDU design, but with enriched fuel, has been put forward as a Generation IV option, along with variants of the various high temperature reactor designs that have been proposed or tested over the years, including the pebble bed helium gas-cooled reactor mentioned earlier, with China taking a lead on that (MacDonald 2016). However, as we shall see, in the main, it has been variants of fast neutron reactors and reactors using molten salts and thorium that have gained most attention, in part since they allegedly offer ways to reduce or recycle wastes. Whether they can also be economic remains to be seen.

The constraints on the process of development of new nuclear technologies are certainly much tighter now than they were in the early days, when nuclear R&D was well funded. It continued to receive the lion's share of R&D funding into the 1970s/1980s, as figure 2.1 illustrates.

Support has tailed off since then, but as Nuttall put it, in the early days '*the engineers were free to design and build technologies at the limits of the possible, without regard to cost, real utility, external impacts or the merits of alternative opportunities*' (Nuttall 2005). Now they have to be more pragmatic. Public energy R&D budgets have been cut worldwide, including for nuclear, and markets for the technologies have toughened, with the Reagan/Thatcher era seeing the emphasis shift to market competition and away from state support and intervention. Private sector energy R&D did not expand to compensate, and the new climate is one of short-termism and tight economic constraints.

As a result of the various accidents and the rise of environmental sensitivity, environmental constraints are also now major factors, as are safety, and, increasingly,

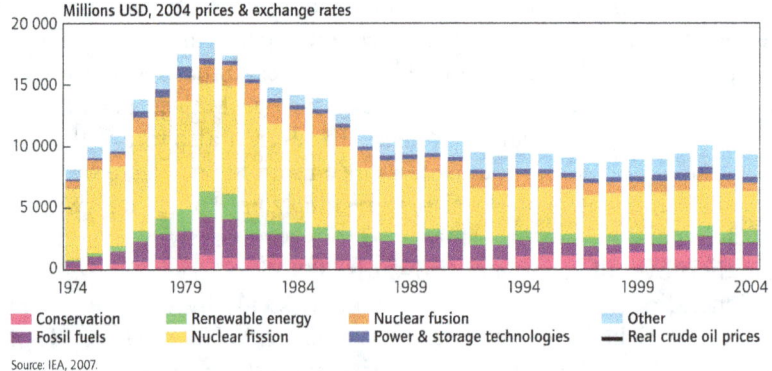

Figure 2.1. Energy R&D funding in the IEA countries (IEA 2007).

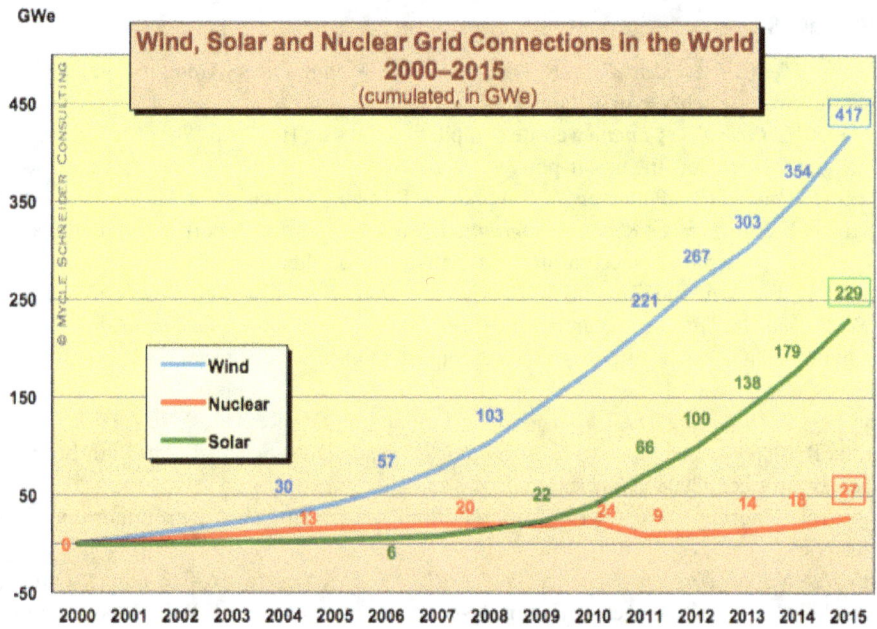

Figure 2.2. Renewable growth accelerates ahead of nuclear (WNISR 2016).

security. Nuclear plants of whatever sort are obvious targets for terrorist attack, either direct, or, potentially, cybernetic (NTI 2016). In addition, nuclear plants will have to integrate into the emerging energy system in which renewables will be increasingly dominant; see figure 2.2. Certainly renewable expansion has outpaced nuclear and that seems set to continue, changing the nature of the energy system. The new nuclear technologies will therefore have to overcome some major challenges, and adapt to the changed situation.

That may require a radical rethink of the overall assessment framework used, for example concerning the risk analysis approach that has been adopted in the past, which has shaped reactor system design and associated regulatory frameworks. A recent study suggested that *'with its failure to plan for the cascade of unexpected beyond design-base accidents, the regulatory emphasis on risk-based probabilistic assessment has proven very limited'* (Dorfman *et al* 2013). In the new energy situation, with variable renewables widely used, there will also be a premium on flexibility, not something nuclear plants can deliver easily.

It will be interesting to see if the Generation IV technologies can meet these challenges. One of the things that has changed since the early days is the advent of new materials, which might make it easier. There certainly is a wide range of technical options (IAEA 2014), developing on the ideas from the earlier phase, and a degree of enthusiasm for developing them.

References

AEC 1961 *The SL-1 Accident Briefing Film report*, US Atomic Energy Commission https://www.youtube.com/watch?v=yjljS0aQbCc

Brook B 2016 *Brave New Climate* website compilation links on IFR https://bravenewclimate.com/integral-fast-reactor-ifr-nuclear-power/

Cochran T, Feiveson H, Patterson W, Pshakin G, Ramana M, Schneider M, Suzuki T and von Hippel F 2010 *Breeder Reactor Programs: History and Status*, Report 8, International Panel on Fissile Materials, Princeton http://fissilematerials.org/library/rr08.pdf

Dorfman P, Fucic A and Thomas S 2013 *Late lessons from Chernobyl, early warnings from Fukushima*, in *Late lessons from early warnings: science, precaution, innovation* European Environment Agency http://www.eea.europa.eu/publications/late-lessons-2/part-c-emerging-issues

Dyson F 1981 *Disturbing The Universe*, Sloan Foundation Science Series, New York

Edwards R 2008 *Nuclear super-fuel gets too hot to handle*, New Scientist, April 9th http://www.robedwards.com/2008/04/nuclear-super-f.html

Energy from Thorium 2016 Energy from Thorium website http://energyfromthorium.com/lftr-overview/

EnergyWatch 2006 *Uranium Resources and Nuclear Energy*, Energy Watch report, Ottobrunn/Aachen http://energywatchgroup.org/wp-content/uploads/2014/02/EWG_Report_Uranium_3-12-2006ms1.pdf

Fuller J 1975 *We almost lost Detroit*, Readers Digest Book, New York.

Gates B 2010 *Innovating to Zero*, Microsoft head, Bill Gates TED talk http://www.ted.com/talks/bill_gates.html

Global Data 2011 *Thorium: The Future Fuel for Nuclear Energy?* power-technology.com web Feature, July 12th http://www.power-technology.com/features/feature123620/?WT.mc_id=WN_Feat

IAEA 2014 *Advances in Reactor Technology for Water-cooled Reactors and Small Modular Reactors* International Atomic Energy Agency Conference presentation by Hadid Subki, September http://www.iaea.org/NuclearPower/Downloadable/News/2014-09-24-nptds/2_NPTDS_Subki_Advanced_Reactors-SMR_r5F.pdf

IEA 2007 *Reviewing R&D Policies*, International Energy Agency, Paris http://www.iea.org/publications/freepublications/publication/reviewing-rd-policies.html

Jokalekar A 2013 *The future of nuclear power: Let a thousand flowers bloom*, Scientific American, December 6th http://blogs.scientificamerican.com/the-curious-wavefunction/2013/12/06/the-future-of-nuclear-power-let-a-thousand-flowers-bloom

MacDonald F 2016 *China says it'll have a meltdown-proof nuclear reactor ready by next year*, Science Alert, February 15th http://www.sciencealert.com/china-says-it-ll-have-a-meltdown-proof-nuclear-reactor-ready-by-next-year

NTI 2016 *Outpacing Cyber Threats: Priorities for Cybersecurity at Nuclear Facilities*, report from Nuclear Threats Initiative, Washington DC http://www.nti.org/analysis/reports/outpacing-cyber-threats-priorities-cybersecurity-nuclear-facilities/

Nuttall W 2005 *Nuclear Renaissance: Technologies and Policies for the Future of Nuclear Power* (Bristol: Institute of Physics Publications)

OECD-NEA 2016 *Uranium 2016: Resources, Production and Demand*, (The Red Book) OECD-Nuclear Energy Agency, Paris http://www.oecd-nea.org/ndd/pubs/2016/7301-uranium-2016.pdf

Ramana M V 2013 quoted in Makhijani, A (2013) *Traveling Wave Reactors: Sodium-cooled Gold at the End of a Nuclear Rainbow?* Institute for Energy and Environmental Research, Takoma Park, Ma http://ieer.org/resource/reports/traveling-wave-reactors-sodium-cooled-gold-at-the-end-of-a-nuclear-rainbow/

Ramana M V 2015 *The Forgotten History of Small Nuclear Reactors*, IEEE Spectrum, April 27th http://spectrum.ieee.org/energy/nuclear/the-forgotten-history-of-small-nuclear-reactors

Ramana M V 2016 *Fast breeder reactors and the slow progress of India's nuclear programme*, Ideas for India, August 16th http://www.ideasforindia.in/article.aspx?article_id=1677

Sim Valley Disaster 2014 *History Channel* documentary https://www.youtube.com/watch?v=vKwK35kKgI8

Sorensen K 2009 *Energy From Thorium: A Nuclear Waste Burning Liquid Salt Thorium Reactor*, Google Tech Talk https://www.youtube.com/watch?v=AZR0UKxNPh8

Sorensen K 2010 *Is Nuclear waste really waste?* Google Tech Talk http://www.youtube.com/watch?v=rv-mFSoZOkE&feature=relmfu

Thorium Energy World 2016a website with embedded Oak Ridge National Lab video *The Molten Salt Reactor Experiment*, 1969 http://www.thoriumenergyworld.com/news/infinite-supply-of-electrical-power-at-low-cost

Till C and Chang Y 2011 *Plentiful Energy -The story of the Integral Fast Reactor*, Create Space, extracts at https://bravenewclimate.com/?s=IFR

Weinberg 2013 *Thorium-Fuelled Molten Salt Reactors*, Report for the All Party Parliamentary Group on Thorium Energy, The Weinberg Foundation, London http://www.the-weinberg-foundation.org/wp-content/uploads/2013/06/Thorium-Fuelled-Molten-Salt-Reactors-Weinberg-Foundation.pdf

Weinberg 2014 Weinberg Foundation, written evidence to the House of Lords Select Committee hearings on the Resilience of the Electricity system, REI0027, September https://www.parliament.uk/documents/lords-committees/science-technology/Resilienceofelectricityinfrasrtucture/Resilienceofelectricityinfrastructureevidence.pdf

WNISR 2016 World Nuclear Industry Status Report, annual independent survey and review, London http://www.worldnuclearreport.org/

IOP Concise Physics

Nuclear Power
Past, present and future
David Elliott

Chapter 3

New brooms: on to Generation IV

Contemporary enthusiasts for what are called Generation IV reactor designs see molten salt thorium reactors and breeders as key options, including possibly small modular variants. There is a flurry of activity in this field, including the involvement of small companies. Some enthusiasts see us as being on the brink of a new nuclear era, led, as in some aspects of space exploration, not so much by unwieldy big corporate projects and government labs, as by small risk-taking startups. However, some of the concepts seem to imply larger projects as well as very advanced technology, although in some cases using old ideas.

There are many new startups and new initiatives in the US, often the result of spin offs from academic research (Eaves 2016). For example, Transatomic involves researchers from MIT, who have developed a uranium-based molten salt reactor concept (Transatomic 2016). NuScale, spun out from Oregon State University, is working on small modular reactors. X-Energy, a new startup company in Maryland, is developing a small high temperature reactor (X-Energy 2017), while Seattle-based Terrapower is developing the Traveling Wave Reactor now backed by Bill Gates, with a 600 MW prototype proposed (Terrapower 2016). More on that below. It is a form of breeder reactor, an idea that is also now back in the race.

3.1 Back to breeders, and thorium

As noted above, there is renewed support for fast neutron reactors. GE Hitachi's proposed *Prism* reactor is one current candidate (GEH 2016). Reactors like this can burn up the plutonium that is in store in the UK and elsewhere, separated out by reprocessing of spent fuel from old uranium reactors, and perhaps more to come from the new uranium reactors. Or, in breeder mode, fast reactors can be used to convert the large existing stockpile of U238/depleted uranium into more plutonium. That can be used in other reactors, including breeders and thorium plants. In addition, fast neutron plants may also be able to burn up some wastes.

These options all come together in the integral fast reactor (IFR) concept, with integral reprocessing (Till and Chang 2011). At present, in some countries, plutonium is extracted from reactor spent fuel via a complex chemical process which has to be carried out elsewhere, in a separate operation; a large reprocessing plant. It is a messy process which gives rise to low and medium activity wastes. To run a significant 'closed cycle' plutonium operation would require many such plants in addition to the breeder plants, with spent and new fuel being transported between them. That opens up many safety and security issues. However, in theory there are ways to do some of this via so called high-temperature pyro-processing, which can, again in theory, perhaps be done near or even within the original reactor complex in parallel with plutonium production. That is the IFR idea. Fully 'closed fuel cycle' systems certainly have their attractions. In theory, a breeder reactor can be run, once the first wave of plutonium has been bred, as a self-sustaining system, without the need for further fuel input apart from U238, i.e. it can then self-generate its own fuel internally.

A variant of this idea is the travelling wave reactor, mentioned above, first conceived of in 1958. In one concept, depleted uranium 238 in a long solid tube is converted to plutonium by neutrons from the fission of a U235 charge at one end, plutonium fission then being self-sustaining and running through the cylinder from one end to the other, in high fuel burn up mode, a little like a burning candle. However, in Terrapower's revision, the fission and fast neutron breeding reaction remains at the centre of a core, fed by fresh fuel from its outer edge with used fuel moving out to the periphery, and liquid sodium being used as a coolant. The breed-burn wave thus does not actually 'travel', instead it is an actively managed 'standing wave', with the fuel being continually moved to get optimal exposure (Ahlfeld *et al* 2011).

As we have seen, experience over the years with fast reactors has been mixed. New US versions, like Prism and the Traveling Wave Reactor, may struggle to do better (Makhijani 2013a, Cochran *et al* 2015). As noted in box 2.1, progress elsewhere is slow. Although Russia, India and China still have ongoing projects, Japan's Monju plant is now to be abandoned.

In a recent critical review of the state of play, Green ended by quoting Allison MacFarlane, former chair of the US Nuclear Regulatory Commission, as saying: *'These turn out to be very expensive technologies to build. Many countries have tried over and over. What is truly impressive is that [...] many governments continue to fund a demonstrably failed technology'* (Green 2016). It will be interesting to see how the new projects fare.

There is also the thorium option, which avoids plutonium production, although it needs some U235 (or another neutron source) to initiate thorium/U233 fission. However, once started, it can in theory be sustained with just thorium being needed subsequently as fuel, and waste burn up may also be possible. While this would be less efficient than in fast neutrons reactors, it is claimed that, with the thorium cycle, there would be less waste to deal with.

Thorium based reactors certainly have their attractions, although it has been argued that some of them have been oversold. For example, a 2010 report from the UK National Nuclear Laboratory was fairly dismissive of the potential for thorium fueled reactors: see box 3.1.

Box 3.1. NNL Thorium Fuel review.

The UK's National Nuclear Labs (NNL) offered a perhaps surprisingly conservative view of the thorium fuel option. It would take years to develop and there were not many benefits, as long as uranium was plentiful. Here is a summary.

Timescale: *'It is estimated that it is likely to take 10 to 15 years of concerted R&D effort and investment before the thorium fuel cycle could be established in current reactors and much longer for any future reactor systems'*. It suggested that *'In the foreseeable future (up to the next 20 years), the only realistic prospect for deploying thorium fuels on a commercial basis would be in existing and new build LWRs (e.g., AP1000 and EPR) or PHWRs (e.g., Candu reactors). Thorium fuel concepts which require first the construction of new reactor types (such as high temperature reactor (HTR), fast reactors and accelerator driven systems (ADS)) are regarded as viable only in the much longer term (of the order of 40+ years minimum) as this is the length of time before these reactors are expected to be designed, built and reach commercial maturity.'*

Recycling: *'While the thorium fuel cycle is theoretically capable of being self-sustainable, this is only achievable with full recycle. This would involve the implementation of THOREX reprocessing and a remote fabrication plant for the U-233 fuel due to the high gamma dose from the feed material, both of which present very large technological, commercial and risk barriers, each with a significant cost component. The use of thorium in place of U-238 as a fertile material in a once-through fuel cycle is much less difficult technically, but only yields a very small benefit over the conventional U-Pu fuel cycle. For example, it is estimated that the approach of using seed-blanket assemblies (the blanket being the surrounding fertile thorium material) in a once-through thorium cycle in PWRs, will only reduce uranium ore demand by 10%. This is considered too marginal to justify investment in the thorium cycle on its own.'*

Proliferation: *'Contrary to that which many proponents of thorium claim, U-233 should be regarded as posing a definite proliferation risk. For a thorium fuel cycle which falls short of a breeding cycle, uranium fuel would always be needed to supplement the fissile material and there will always be significant (though reduced) plutonium production.'*

Economics: *'NNL believes that while economic benefits are theoretically achievable by using thorium fuels, in current market conditions the position is marginal and insufficient to justify major investment. There is only a very weak technical basis for claims that thorium concepts using seed-blanket PWR cores will be economically advantageous. The only exception is in a postulated market environment of restricted uranium ore availability and thus very high uranium prices. This is not considered very likely for the foreseeable future, given that economically recoverable uranium reserves are thought to be very price dependent and therefore if uranium prices were to increase, then more uranium would be available to the market.'*

Radiotoxicity: *'Claims that thorium fuels give a reduction in radiotoxicity are justified. However, caution is required because many such claims cite studies based on a self-sustaining thorium cycle in equilibrium. More realistic studies which take account of the effect of U-235 or Pu-239 seed fuels required to breed the U-233 suggest the benefits are more modest. NNL's view is therefore that thorium fuel cycles are likely to offer modest reductions in radiotoxicity. It is considered that the realistic benefits are likely to be too marginal to justify investment in the thorium fuel cycle. However, the substantial reduction in radiotoxicity promised by a full thorium recycle does provide a significant incentive in the long term'* (NNL 2010).

However, this study did not look at thorium used in molten salt reactors, and enthusiasts argue that MSRs have some added advantages over reactors using solid fuel. The Weinberg Foundation says they *'have lower fissile inventories, no radiation damage constraint on fuel burn-up, no spent nuclear fuel, no requirement to fabricate and handle solid fuel, and a homogeneous isotopic composition of fuel in the reactor'* (Weinberg 2016).

The liquid fluoride thorium reactor (LFTR) variant is a 'two-fluid' design. Fertile thorium in an outer 'blanket' is separated from fissile U235/Pu239 (and U233) in an inner container core made of graphite, which acts as a moderator. Neutrons from fissions in the core create U233 in the blanket which is then fed to the core to undergo fission. All these materials are in molten salt form. Fission in the molten salts in the core is the heat source and these salts are cycled through a heat exchanger to extract it—there is no need for other cooling systems.

Fluoride salt mixtures are chemically stable and impervious to radiation damage and have good heat retention capacity. As the *Energy from Thorium* website notes, they can dissolve useful quantities of actinide fluorides such as uranium tetrafluoride, thorium tetrafluoride and plutonium trifluoride, and, in a reactor context, they can chemically capture fission products such as cesium and strontium in fluoride form and prevent their release. Moreover, although LFTRs would run at high temperatures, they would not need high pressures, so avoiding decompression risks. It is also claimed that they would produce fewer wastes than conventional uranium-based systems. However, although it is all done in a fluid matrix, it is quite a complex system in terms of internal plumbing and materials handing: see box 3.2.

Box 3.2. LFTR fuel cycle.

In LFTRs, U233 produced in the outer blanket has to be moved to the core and more thorium added to the blanket. Fresh uranium (or Pu239) also has to be added to the core and wastes removed. The *Energy from Thorium* website describes these processing/transfer stages as follows: *'Fission of U-233 in the reactor generates thermal power as well as excess neutrons that would be captured in a blanket fluid containing thorium tetrafluoride in solution. Thorium, having absorbed a neutron, first decays to protactinium and ultimately to uranium-233. New fuel would be chemically removed from the blanket fluid either at the uranium stage or the protactinium stage, which has additional complexity and advantages. The new uranium fuel would be introduced into the fuel salt of the LFTR at the same rate at which it is consumed. Uranium-233 is consumed at high efficiency (91%) in a thermal spectrum reactor and that which is not consumed goes on to form uranium-235, which is also consumed at high efficiency (85%) in a thermal spectrum reactor. This limits considerably the amount of material that can reach the stage of the first transuranic, in this case, neptunium-237, and thus the issue of long-lived actinide waste production. The fuel salt used in the LFTR is chemically processed as the reactor operates, removing fission products while retaining actinide fuels. This allows the creation of an actinide-free waste stream which decays to acceptable radioactivity levels in approximately 300 years'* (Energy from Thorium 2016).

> In terms of safety, the Weinberg Foundation says unlike with reactors using solid fuel, *'as fuel in a LFTR is already in liquid form, it cannot melt down [...] The LFTR can only overheat if the circulation of the molten salt is disrupted as a result of a loss of power, thereby preventing heat removal from the core. If that should happen, the build-up of excess heat melts a freeze plug of solidified salt at the bottom of the reactor, allowing the molten salt to drain into a separate tank. Once in the tank, the fission reaction stops and the liquid fuel cools down and becomes an inert solid mass. This is an entirely passive process requiring no external power source.'* (Weinberg 2012).

While it is claimed that LFTRs have higher inherent safety features and fewer technical obstacles and waste production problems than uranium reactors, there are some uncertainties. When non-fissile thorium is converted into fissile U233, there are decay products, starting with U232, which are very radioactive, much more so that U235 or plutonium, emitting hard gamma radiation. That makes the fuel difficult to work with, whether in solid or liquid form, and means much more shielding as well as even tighter protection against leaks.

It is claimed that, given the high U233 burn up and the reduced by-products from its fission, thorium reactors produce less wastes. However, to get U233, initially U235 and/or Pu 239 must be used, either in a separate operation, or (as in the LFTR above), in the molten salt. So there will still be some waste to be dealt with. The Weinberg Foundation says that *'17% of fission products from an LFTR have long half-lives and these will require safe storage for up to 300 years'*. If actually achieved, that would be an improvement on conventional uranium fission plants, but it is still an issue.

It has been suggested that capital costs for MSRs could be around half that for conventional nuclear plants (Hargraves and Moir 2011). That is very speculative, even assuming mass production, but the prospect of lower costs has driven interest in MSRs, as has the inherent safety features that are claimed. In theory, the use of molten fluoride salt would allow for the reaction to be self-limiting. The Weinberg Foundation says that *'MSR temperature regulation is passive, so no control rods or active cooling system are required. The heat generated by fission expands the molten salt, decreasing the level of reactivity, which leads to contraction of the molten salt and an increase in reactivity, and thus a self-regulating system'*. In addition, *'the passive temperature system enables MSRs to load-follow automatically. As more heat is extracted from the reactor, the salt temperature goes down, causing power output to increase, thus responding instantaneously to demand. In addition, MSRs have far greater load-following capability than solid-fuelled reactors due to their capacity for online removal of xenon gas. Xenon is a neutron-poison which increases in quantity in solid-fuelled reactors when the power level is lowered, thus limiting load-following operation. In MSRs, xenon is constantly bubbled out of the reactor via the off-gas system which enables MSRs to be among the most flexible load-following nuclear reactors'* (Weinberg 2014).

> **Box 3.3. Current thorium reactor projects.**
>
> India has significant thorium reserves and there have been constraints on the import of uranium since it has not signed the Non-Proliferation Treaty. It is seeking to develop a thorium-based advanced heavy water reactor (AHWR). *World Nuclear News* noted: *'The long-term goal of India's nuclear programme has been to develop an advanced heavy-water thorium cycle. The first stage of this employs pressurized heavy-water reactors and light water reactors, to produce plutonium. Stage two uses fast neutron reactors to burn the plutonium and breed uranium-233 from locally mined thorium. The blanket around the core will have uranium as well as thorium, so that further plutonium is produced as well. In stage three, AHWRs burn the uranium-233 from stage two with plutonium and thorium, getting about two thirds of their power from the thorium'* (WNN 2009). However, this ambitious programme has proven slow to get started, but India is certainly trying hard on thorium: with the help of the US, it also aims to build the world's first thorium accelerator driven system (Thorium Energy World 2016b).
>
> In 2011 China's Academy of Sciences said it had chosen a *'thorium-based molten salt reactor system'* for further research, in a $350 million programme, and work on that has continued (Duggan 2014). France has had an MSR R&D programme since 1997, including work on fast neutron variants of MSR designs, and private companies in the US, Japan, Australia and the Czech Republic are developing thorium-fuelled MSR designs. Russia is also looking at MSRs, as is Norway, which has thorium reserves. However, the largest reserves are in Australia. But worryingly, some thorium-derived U233 from the early US programme is evidently missing (Alvarez 2014).

Xenon poisoning and flash off were key factors in the Chernobyl disaster, so you would need to be very sure this new untried concept really worked. Apart from that and the radiation from U232 decay, the safety case has been seen as relatively good, compared to standard reactors (Elsheikh 2013). Proliferation resistance is also said to be good (no plutonium is produced directly), but there are some security issues. U233 (with its U232 content) is hard to work with, but it can be used to make weapons, so thorium reactors can have implications for weapons proliferation (Phys. org 2012). Moreover, in terms of hardware, although there are projects in India, China and elsewhere (see box 3.3), there is still some way to go before the thorium option is proven, with fuel management and waste burn up issues still to be resolved.

In addition to waste burn up in IFRs and MSRs, there is the waste transmutation option, using particle beams, sometimes called ADS. Carlo Rubbia at the CERN Laboratory in Switzerland had suggested firing a beam of neutrons into a sub-critical mix of thorium and plutonium, the neutron flux converting wastes into less hazardous forms.

The energy output is seen as a by-product. It is claimed to be fail-safe, since fission could be instantly halted by switching off the accelerator. However, there are issues: only some wastes would be suited to transmutation and achieving significant conversion rates would probably need a multi-stage process, with repeated reprocessing steps to separate out target isotopes for retreatment. Moreover, at each stage, new wastes would be created (NEA 2002).

There is nevertheless still some interest in ADS, perhaps as a new Generation IV option. For example, the EMMA system (electron model for many applications) being developed by researchers at the UK's Daresbury Labs in Cheshire. It would bombard thorium, and (some) nuclear wastes, in a sub-critical reactor, producing heat, which can be used to generate power. Essentially it is a development of Rubbia's idea (Burns 2012).

However, in general, it has been more conventional generation options, such as molten salt-based systems, and the development of *small reactors*, that has captured most attention. The next section looks at what has been done so far in this context and at plans for the future.

3.2 Small is beautiful

Small modular reactors (SMRs) of up to a few hundred megawatts capacity are being touted by some as the way ahead for nuclear power since they are expected to be quicker to build than large gigawatt scale plants and so less costly to finance, with mass production reducing unit costs (NNL 2012). They might also be located in or near cities, so that the waste heat they produce could be fed to district heating networks, the sale of this extra output offsetting their cost further. It may also be possible, it is claimed, that their power output could be more easily varied, so that they could play a role in grid balancing, though, as with large nuclear plants, that would undermine their economics—it is best to run them flat out.

There has certainly been much enthusiasm expressed for SMRs, with various vendors offering their wares, for example from Westinghouse (Westinghouse 2016). Estimates of costs vary, but, in the US context, it has been claimed by NuScale, a company spun out of Oregon State University, that 'first of kind' SMRs might generate at around 101 c MWh^{-1}, falling to 90 c MWh^{-1} on mass production, cheaper than new large nuclear plants, at 96 c MWh^{-1}, and also cheaper than coal fired plants, but not competitive with unabated combined cycle gas plants (64–66 c kWh^{-1}) or wind plants (80 c kWh^{-1}) or hydro (85 c kWh^{-1}) (NuScale 2016).

NuScale seems to be leading the pack with its water-cooled design. Terrestrial Energy amongst others are pushing the less developed molten salt reactor idea (WNN 2016a), but it may take some time to get to a commercial project. However, more conventional designs may get going sooner: the UKs Energy Technologies Institute said SMRs could be running in the UK by 2030 if R&D work gets underway soon, with heat supply an extra option (ETI 2016).

Although much work will have to be done to modify the technology for civil power (and possibly heat), it is sometimes claimed that civil SMRs can be based on existing nuclear submarine propulsion technology, which is well established, with companies like Rolls Royce being well placed to develop suitable units (Tovey 2016). However, as was noted earlier, the submarine and civil contexts are very different, with different operational requirements and regimes. Safety and reliability is obviously a key issue in all contexts, but even in the closely managed military environment things can go wrong (Edwards 2011). And spreading SMRs around in urban areas could pose safety and security risks, with local public acceptability potentially being a major issue. Reactors are usually sited well away from cities.

In the US however, the Tennessee Valley Authority (TVA) claims that SMRs could be put close to population zones with a reduced need for extensive emergency planning evacuation zones. They say that, given safety upgrades, *'based upon the preliminary information which we've received from the four vendors, we are confident that all of them can be supported by a two-mile emergency plan [zone] and at least one has capability of site [only] boundary'*, i.e. no safety zoning beyond the plant site. That compares with 10 miles typically required for large reactors (Nuclear Energy Insider 2016).

That seems a little provocative. Will anyone accept mini-nukes in their backyards? And what about security? SMRs will presumably be sealed modular units, making access to the fissile material hard, but, unless they are very carefully guarded, they might still provide an enticing and convenient target for terrorist attack. Once the fissile context was exhausted, the cores or the whole power unit would also have to be taken somewhere for spent fuel extraction and recharging, assuming that could not be done on site. Either way, there are security issues.

In terms of operational safety, the US Union for Concerned Scientists says: *'Multiple SMRs may actually present a higher risk than a single large reactor, especially if plant owners try to cut costs by reducing support staff or safety equipment per reactor'*. It adds that *'some proponents have suggested siting SMRs underground as a safety measure. However, underground siting is a double-edged sword—it reduces risk in some situations (such as earthquake) and increases it in others (such as flooding). It can also make emergency intervention more difficult. And it increases cost'* (UCS 2013).

There are thus a range of technical, economic, safety and security issues to face, with there being no clear indication that they can be resolved (Ramana and Mian 2014); see box 3.4.

Box 3.4. SMR design conflicts.

Academics M V Ramana and Zia Mian note: *'There are a very wide variety of SMR designs with distinct characteristics'* which *'vary by power output, physical size, fuel geometry, fuel type and enrichment level (and resulting spent fuel isotopic composition), refueling frequency, site location, and status of development'*. Following Alexander Glaser of Princeton University, they categorise them in four families. The first involves reactor designs intended to be the first on the market, essentially scaled-down standard PWRs typically fuelled with low enriched uranium, 5% or less. The second family involves the high temperature gas-cooled reactors, typically using uranium enriched to well above 5%, graphite as a moderator, and helium as the coolant. The third category use fast neutrons without any moderator possibly with sodium coolant, running on and generating plutonium. The fourth family are essentially 'nuclear batteries', with long-lived cores and highly enriched fuel, designed for possibly unattended operation, perhaps in remote locations in developed countries.

Raman and Mian have identified a range of potential design criteria conflicts with these various SMR concepts: *'Proponents of the development and large scale deployment of small modular reactors suggest that this approach to nuclear power technology*

> *and fuel cycles can resolve the four key problems facing nuclear power today: costs, safety, waste, and proliferation. Nuclear developers and vendors seek to encode as many if not all of these priorities into the designs of their specific nuclear reactor. The technical reality, however, is that each of these priorities can drive the requirements on the reactor design in different, sometimes opposing, directions. Of the different major SMR designs under development, it seems none meets all four of these challenges simultaneously. In most, if not all designs, it is likely that addressing one of the four problems will involve choices that make one or more of the other problems worse.'* (Ramana and Mian 2014)

There certainly has been no shortage of critical analysis of SMR's technical viability, safety, security and economics in the US (Makhijani 2013b, Cooper 2014) and a fairly downbeat analysis from the US government Accountability Office (GAO 2015). The issue of public acceptability has also featured in the debate in the UK (MacKerron 2015). So have the broader strategic issues: the non-nuclear alternatives are arguably more credible (Toke 2016). That is a view echoed in some US coverage (Hyman and Tilles 2016).

From the industry side however, enthusiasm remains strong, with much debate about exactly how to proceed and on what basis. For example, should SMRs replace large nuclear plants in any future programme? In the UK context, ETI's 2015 report 'The role for nuclear within a low carbon energy system' said that, while nuclear plants were best used for continuous base-load supply, there was a limited number of sites where new large plants might be installed, mainly existing nuclear sites, where opposition should be less, with *'an upper capacity limit in England and Wales to 2050 from site availability of around 35 GW'*. However, there could be room for at least 21 GW SMRs in the UK, given that more sites could be available for them, including near cities, where the heat option offered an economic compensation.

So, although SMRs *'may be less cost effective for baseload electricity production, SMRs could fulfil an additional role in a UK low carbon energy system by delivering combined heat and power (CHP)—a major contribution to the decarbonisation of energy use in buildings'*. That assumed the necessary district heating infrastructure was available, with SMRs delivering heat into cities *'via hot water pipelines up to 30 km in length'* (ETI 2015).

An SMR assessment programme has been launched by the UK government, and SMR programmes are going ahead in the US and elsewhere (WNA 2016). This is all in a context where, according to the UK National Nuclear Laboratory, the potential market for SMRs might be up to 85 GW by 2035 (WNN 2016b). That may be optimistic. The OECD says *'up to 21 GW of SMRs could be added globally by 2035'* (NEA 2016). Interestingly, at least in the UK context, SMRs seem to be seen as *additional* to conventional nuclear plants, with *Penultimate Power* saying SMRs will work best as *'complementary to, rather than competing with'* large-scale nuclear plants (Muccusker 2016).

The current flurry of enthusiasm for SMRs seems to be mainly driven by the failure of conventional nuclear to expand as fast as the industry would like.

However, given the potential limits on site acceptability, SMRs near or in cities may not offer much help in changing this situation, although they may boom if there are technological breakthroughs, or if a wider market emerges (Larson 2016). For example, some projects may be suited to specialised applications, such as supporting industrial processes (Carlsson *et al* 2012).

Or they may find uses as power sources in remote areas, perhaps with easy-to-deploy mobile units, possibly sited offshore, floating, or even underwater, as in the French Flexblue seabed concept (WNN 2011). MSRs are said not to need water to operate, but most other SMR designs will presumably need some for cooling, so this type of location has attractions.

For the moment, while there are several prototype projects underway and many more being considered in many countries, most of them are just concepts (see box 3.5). Some of these projects may eventually succeed technically. However, to succeed economically, there would have to be orders for large numbers, so as to obtain the hoped-for economies of mass production of identical factory-made units. Given the likely constraints on local acceptability, that may not happen. In which case, SMRs may have a limited role. In the past, the nuclear industry has tried to improve the economics of nuclear plants by going for larger plants, without too much success. It is not clear if small plants will have any more success. Scaling down does not necessarily reduce complexity, and it is that which may drive costs most.

Steve Kidd, a nuclear industry lobbyist, says that *'the jury is still out on SMRs'* and on whether they are *'a viable solution to some of the problems experienced by projects to build large light water reactors (LWRs)'*. Assuming they are technically viable, *'the smaller capital expenditure needed to build a largely factory-built smaller unit and the shorter construction period are certainly attractive features. And if*

Box 3.5. SMR design concept families. (Roulstone 2015)

Water-cooled SMRs CAREM-25 (Argentina) ACP100 (China) Flexblue (France) AHWR300 (India) IRIS (International) DMS (Japan) IMR (Japan) SMART (S Korea) KLT-40S (Russia) VBER-300 (Russia) ABV-6M (Russia) RITM-200 (Russia) VVER300 (Russia) VK-300 (Russia) UNITHERM (Russia) RUTA-70 (Russia) mPower (US) NuScale (US) Westinghouse SMR (US) SMR-160 (US) Elena (Russia) SHELF (Russia)

High temperature gas-cooled SMRs HTR-PM (China) GTHTR300 (Japan) GT-MHR (Russia) MHR-T (Russia) MHR-100 (Russia) PBMR-400 (SA) HTMR-100 (SA) EM2 (US) SC-HTGR (US) Xe-100 (US) U-Battery (UK)

Liquid-metal cooled fast SMRs CEFR (China) PFBR-500 (India) 4S (Japan) SVBR-100 (Russia) BREST-300 (Russia) PRISM (US) Gen4 Module (US) Astrid (France)

Molten-salt cooled SMRs Terrestrial En (Canada) Seaborg Tech (Den) Fuji (Japan) LFTR (China) Moltex (UK) EVOL (EU) Flibe Energy (US) WAMSR Transatom (US).

electricity production is moving away from large centralised generating units into a distributed power model, smaller nuclear units may still have a chance'.

However, he takes the traditionalist view that big is best: *'Smaller nuclear reactors were developed back in the 1950s but the sensible decision was made to take advantage of nuclear's real unique selling proposition. That is the ability to produce huge quantities of electricity very reliably in one place, with a small fuel input and minimal environmental impact. Reactor units became progressively larger in an attempt to capture economies of scale in construction costs, but also (and very importantly) to minimise operating and maintenance (O&M) expenses'.*

He accepts that SMRs *'may have a chance today in remote areas in developed countries that don't have easy grid access',* but concludes that, elsewhere, *'unless the regulatory system in potential markets can be adapted to make their construction and operation much cheaper than for large LWRs, they are unlikely to become more than a niche product. Even if the costs of construction can be cut with series production, the potential O&M costs are a concern. A substantial part of these are fixed, irrespective of the size of reactor'* (Kidd 2015).

He is not alone in being uncertain about SMRs. The UK's Lloyds Register, in an otherwise optimistic review of nuclear (and solar) prospects, said the potential contribution of SMRs *'is unclear at this stage'* and *'its impact will most likely apply to smaller grids and isolated markets',* with some contributors to the review saying that, in general, SMRs had a *'low likelihood of eventual take-up'* (Lloyd's Register 2017).

3.3 Reactor choices and progress

As we have seen, one of the choices that emerged in the early days of nuclear development was between plutonium breeders and thorium molten salt reactors, with, amongst others, nuclear pioneer Alvin Weinberg a champion of the latter against the then dominant view that plutonium breeders were the way ahead. He lost. However, we are now back at a stage when the choice may be revisited, although the range of choices is now more complex.

As was noted in box 2.2, there may be a basic choice to be made between systems in which plutonium is used and more bred for later use, or thorium is converted to U233, using fast neutrons, with neutron absorption optimised, as against systems which just use uranium and 'thermalised' slow neutrons, optimising fission, with less breeding. Molten salt reactors can in theory be run either way, but it is easier to breed U233 from thorium with fast neutrons. MSRs can be self-cooled but most breeders use liquid sodium, which has clear risks, while high temperature reactors (HTRs), which also have proponents, use helium gas cooling. The pebble bed reactor being developed in China and the US Xe-100 variant are examples.

Transmutation of active wastes is possible in all types of reactor, although the absorption potential is higher with fast neutrons. Scale may also not be the defining technical factor. Plutonium/U233 breeding in sodium-cooled breeder reactors like IFRs or in molten salt reactors, like the LFTR, is in theory possible at any scale. Even so, there may be a choice ahead between large IFR-type plants, including

Prism type burners or breeders, and small plants, SMRs of various kinds, including LFTRs, whether thermal or fast, and HTRs.

The debate over reactor types, fuels and scales can get both tortuous and lively (Green 2015). On one hand there are those who want to stick with what is known and certainly there has been massive investment in uranium-based nuclear technology and associated infrastructure.

On the other hand, some think uranium reserves will not be sufficient for the longer term and that, in any case, the existing technology is not delivering competitive energy: new technology is needed. So then we move into the debate over which option or options to choose. There are devotees of each. Some look to plutonium breeders, since they can use U238/depleted uranium, of which there is plenty. They may also be able to burn up waste. Others look to thorium-based reactors in the belief that they will produce less wastes than plutonium breeders and can use a fuel that is more abundant than uranium. The latter is certainly true, but since thorium is not fissile and has to be converted to U233, the resource comparison should perhaps be between fertile U238 and thorium (Dracoulis 2011).

The waste production issue is more complex: the profile of what would be produced by thorium molten salt plants would be different, with fewer long lived wastes than with plutonium breeders, and it may be easier to extract and process. It may also be possible to burn some of it up directly. Though there will still be some wastes and proliferation issues still exist, as they do (even more so) with breeders, unless IFR-type burn-up works well.

Whichever system or systems are selected, it will take time to develop commercial scale plants, or even prototypes. For example, in an interesting exchange of views in a debate on thorium run by *The Engineer*, Fiona Rayment, director of fuel cycle solutions at the UK's National Nuclear Laboratory, said: *'To develop radical new reactor designs, specifically designed around thorium, would take at least 30 years'* (Harris 2014).

She may be proved wrong, or other options may step in, including IFRs and HTRs, but it is early days as yet, with, on the one hand, plenty of room for all types and scales, while on the other, it being unclear whether any of these ideas will prosper beyond prototypes. To some extent, the Generation IV technologies seem to be at the same stage as many renewables were twenty or more years ago, seeking to step beyond some historical precedents. Renewables of various types have managed that. It remains to be seen if Generation IV can.

Meantime, box 3.6 provides some 2014 estimates of hoped for progress, produced by the Generation IV International Forum (GIF). On the basis of the industry's past performance, they may be optimistic. GIF noted that the Fukushima disaster had led to some slow downs.

Certainly, even on these estimates, there is still a long way to go before commercialisation. GIF's 2014 update to its 2002 Technology Roadmap review noted that *'the development of technologies and associated system designs to the point of commercialisation for each of the six systems, as identified in the original Technology Roadmap, would have required considerable investment and international*

> **Box 3.6. GIF Generation IV progress estimates. (GIF 2014)**
>
>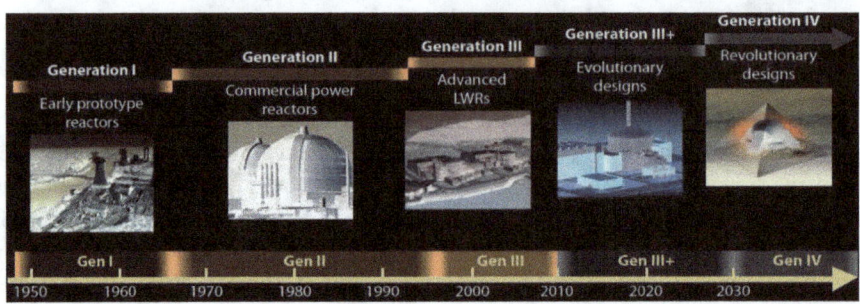
>
> **Figure 3.1.** GIF's depiction of succeeding Nuclear Reactor Generations I-V (GIF 2014).
>
> The Generation IV International Forum 'Technology Roadmap Update for Generation IV Nuclear Energy Systems' measures progress according to three (pre-commercialisation) phases:
> - The viability phase, when basic concepts are tested under relevant conditions and all potential technical show-stoppers are identified and resolved;
> - The performance phase, when engineering-scale processes, phenomena and materials capabilities are verified and optimised under prototypical conditions; and
> - The demonstration phase, when detailed design is completed and licensing, construction and operation of the system are carried out, with the aim of bringing it to commercial deployment stage.
>
> Its 2014 projections were as below, with demonstration phases presumably to follow on after:
> - *Gas-cooled fast reactor*: end of viability phase 2022; end of performance phase 2030.
> - *Molten salt reactor*: end of viability phase 2025; end of performance phase 2030.
> - *Sodium-cooled fast reactor*: end of viability phase 2012; end of performance phase 2022.
> - *Supercritical-water-cooled reactor*: end of viability phase 2015; end of performance phase 2025.
> - *Very-high-temperature reactor*: end of viability phase 2010; end of performance phase 2025.
> - *Lead-cooled fast reactor*: end of viability phase 2013; end of performance phase 2021.

commitment. Since the "starting point" and R&D funding of the different Generation IV systems were not equivalent, the degree of technical progress over the past decade has not been uniform for all systems'.

So development is uneven: '*A number of participating countries devoted significant resources to the development of the SFR and VHTR, for example, in large part due to the considerable historical effort associated with these technologies. More limited resources were dedicated to the other systems'* (GIF 2014).

A more recent review of advanced reactor options, from a US perspective, carried out by the Argonne, Idaho and Oak Ridge National Labs, looked at basically the same six Generation IV advanced reactor technology concepts. It concluded, a little more optimistically, that *'the modular high temperature gas-cooled reactor (HTGR) and sodium-cooled fast reactor (SFR) have high enough technology readiness levels to support a commercial demonstration in the near future. These technologies are considered mature as a result of several successful demonstrations brought about through billions of dollars of public and private investment in the U.S. over more than fifty years. These systems are also being built internationally, further confirming the high level of maturity of these systems as evaluated in this study.'*

However, *'the fluoride-cooled high-temperature reactor (FHR) and lead-cooled fast reactor (LFR) are less mature and require additional research and development and engineering demonstration in the near future. International and U.S. technology development activities are underway to mature these technologies, and technology demonstrations are planned. Other options examined (e.g., gas-cooled fast reactor) were even lower in maturity or did not have significant U.S. commercial interest (e.g., super-critical water-cooled reactor)'.*

On this basis it suggested that HTGRs and SFRs *'are mature enough to enable deployment of their first modules at commercial scale (the commercial demonstration step) in the early 2030s with additional commercial offerings soon thereafter',* while the less-mature FHR and LFR *'are facing a longer technology development path to commercial offerings because they need a combination of both the engineering demonstration step and the performance demonstration step through 2040, prior to commercial offerings in ~2050'* (INL 2017).

It will be interesting to see if any of these quite long-term predictions prove to be correct.

The optimists look to rapid development of prototypes: in January 2017 NuScale applied for NRC clearance for a 12 module SMR test plant and it has been reported that by late 2019 Terrestrial Energy may be ready to apply for a test of its more complex MSR design (WNN 2017a, 2017b). However, as we have seen, not everyone is optimistic about the potential for rapid or significant gains. In a recent interview-based review of advanced nuclear prospects in North America, Eaves looks at both sides of the debate, reporting on the enthusiasm expressed by some practitioners, but she also quotes former US Nuclear Regulatory Commission Chairman Allison Macfarlane as saying: *'I do not see past experience pointing at a positive direction'* (Eaves 2016).

Macfarlane was talking about high temperature reactors, which some think could be the leader, but as we have seen, that comment might also be applicable, in the view of some critics, to most of the new nuclear options. Certainly, there are multiple challenges. Breakthroughs are always possible, and as one of the nuclear enthusiasts that Eaves interviewed said: *'If we believe that nothing new can happen and everything is really hard, then it will be. That's not to minimize the challenge, but it is to say, if you start out thinking something is impossible, it's very unlikely to happen'.* However, given the past history of nuclear power, and the multiple challenges, a degree of caution seems wise.

A group of 10 energy-specialist academics recently contributed a series of comments to a wide-ranging review of the 'Frontiers of Energy' in *Nature Energy*. While recognising the technical, social and environmental challenges ahead, most were optimistic about the role that renewables and energy efficiency could play in response to climate change and the other energy-related issue that we face. By contrast, the entry on nuclear power (by Ramana) concluded that *'no reactor design seems capable of simultaneously overcoming all the challenges confronting nuclear power. Besides economics and safety, these also include the generation of radioactive waste, the linkage to nuclear weapons, and the consequent public opposition'* (Armstrong *et al* 2016). That is pessimistic, but it may be right.

References

Ahlfeld C *et al* 2011 *Conceptual Design of a 500 MWe Traveling Wave Demonstration Reactor Plant*, Proceedings of ICAPP 2011, Nice, France, May 2–5, paper 11199 http://terrapower.com/uploads/docs/ICAPP_2011_Paper_11199.pdf

Alvarez R 2014 *Thorium: the wonder fuel that wasn't*, Bulletin of the Atomic Scientists, May 11th http://thebulletin.org/thorium-wonder-fuel-wasnt7156

Armstrong R, Wolfram C, de Jong K, Gross R, Lewis N, Boardman B, Ragauskas A, Ehrhardt-Martinez K, Crabtree G and Ramana M V 2016 *The Frontiers of Energy*, Nature Energy 1:15020 http://www.nature.com/articles/nenergy201520

Burns C 2012 *What do you get when you cross an accelerator with a nuclear reactor?*, The Guardian, February 9th https://www.theguardian.com/science/blog/2012/feb/09/accelerator-nuclear-reactor

Carlsson J, Shropshire D, van Heek A and Fütterer M 2012 Economic viability of small nuclear reactors in future European cogeneration markets *Energy Policy* 43 396–406 http://www.sciencedirect.com/science/journal/03014215/43/supp/C

Cochran T, Feiveson H, Mian Z G, Ramana M, Schneider M and von Hippel F 2015 It's Time to Give Up on Breeder Reactors, *Bulletin of the Atomic Scientists* 66 Issue 3 http://www.tandfonline.com/doi/10.2968/066003007

Cooper M 2014 *The Economic Failure of Nuclear Power and the Development of a Low Carbon Electricity Future: Why Small Modular Reactors are part of the problem, not the solution*, Institute for Energy and the Environment Vermont Law School, May http://www.nirs.org/reactorwatch/newreactors/cooper-smrsaretheproblemnotthesolution.pdf

Dracoulis G 2011 *Thorium is no silver bullet when it comes to nuclear energy, but it could play a role*, The Conversation, August 5th https://theconversation.com/thorium-is-no-silver-bullet-when-it-comes-to-nuclear-energy-but-it-could-play-a-role-1842

Duggan J 2014 *China working on uranium-free nuclear plants in attempt to combat smog*, The Guardian, March 19th https://www.theguardian.com/world/2014/mar/19/china-uranium-nuclear-plants-smog-thorium

Eaves E 2016 Can North America's advanced nuclear reactor companies help save the planet? *Bulletin Of The Atomic Scientists* 73 http://www.tandfonline.com/doi/full/10.1080/00963402.2016.1265353

Edwards R 2011 *Flaws in nuclear submarine reactors could be fatal, secret report warns*, The Guardian, March 10th http://www.theguardian.com/world/2011/mar/10/royal-navy-nuclear-submarine-reactor-flaws

Elsheikh B 2013 Safety assessment of molten salt reactors in comparison with light water reactors, *Journal of Radiation Research and Applied Sciences* **6** Issue 2 63–70 http://www.sciencedirect.com/science/article/pii/S1687850713000101

Energy from Thorium 2016 Energy from Thorium website http://energyfromthorium.com/lftr-overview/

ETI 2015 *The role for nuclear within a low carbon energy system*, Energy Technologies Institute https://s3-eu-west-1.amazonaws.com/assets.eti.co.uk/legacyUploads/2015/09/3511-ETI-Nuclear-Insights-Lores-AW.pdf

ETI 2016 *Preparing for deployment of a UK small modular reactor by 2030*, Energy Technologies Institute http://www.eti.co.uk/library/preparing-for-deployment-of-a-uk-small-modular-reactor-by-2030

GAO 2015 *Nuclear Reactors: Status and challenges in development and deployment of new commercial concepts*, US Government Accountability Office, July, GAO-15-652 http://www.gao.gov/assets/680/671686.pdf

GEH 2016 Prism reactor, GE Hitachi website http://gehitachiprism.com/what-is-prism/how-prism-works/

GIF 2014 *Technology Roadmap Update for Generation IV Nuclear Energy Systems*, Generation IV International Forum (GIF) http://www.gen-4.org/gif/jcms/c_60729/technology-roadmap-update-2013

Green J 2015 *Thor-bores and uro-sceptics: thorium's friendly fire*, Nuclear Monitor, April 9th https://www.wiseinternational.org/nuclear-monitor/801/thor-bores-and-uro-sceptics-thoriums-friendly-fire

Green J 2016 *Japan abandons Monju fast reactor: the slow death of a nuclear dream*, The Ecologist, October 6th http://www.theecologist.org/News/news_analysis/2988203/japan_abandons_monju_fast_reactor_the_slow_death_of_a_nuclear_dream.html

Hargraves R and Moir R 2011 Liquid Fluoride Thorium Reactors *Physics and Society* **40** 9 https://www.aps.org/units/fps/newsletters/201101/upload/january2011.pdf

Harris S 2014 *Your questions answered: thorium-powered nuclear*, The Engineer, January 9th https://www.theengineer.co.uk/issues/december-digital-edition-2/your-questions-answered-thorium-powered-nuclear/

Hyman L and Tilles W 2016 *Why Small Modular Reactors Are Not The Next Big Thing*, Oilprice.com, April 7th http://oilprice.com/Alternative-Energy/Nuclear-Power/Why-Small-Modular-Reactors-Are-Not-The-Next-Big-Thing.html

INL 2017 *Advanced Demonstration and Test Reactor Options Study*, Argonne, Idaho and Oak Ridge National Labs report: https://art.inl.gov/INL%20ART%20TDO%20Documents/Advanced%20Demonstration%20and%20Test%20%20Reactor%20Options%20Study/ADTR_Options_Study_Rev3.pdf

Kidd S 2015 *Nuclear myths - is the industry also guilty?*, Nuclear Engineering International, June 11th http://www.neimagazine.com/opinion/opinionnuclear-myths-is-the-industry-also-guilty-4598343

Larson A 2016 *Is There a Market for Small Modular Reactors?* Power magazine, January 6th http://www.powermag.com/market-small-modular-reactors/

Lloyd's Register 2017 Technology Radar reviews: 'A Nuclear Perspective', and 'Low Carbon report' (London: Lloyd Register) www.lr.org/en/low-carbon-power/technology-radar.aspx

MacKerron G 2015 Small Modular Reactors—a real prospect? Sussex Energy Group Blog, University of Sussex, October 9th http://blogs.sussex.ac.uk/sussexenergygroup/2015/10/09/small-modular-reactors-a-real-prospect-by-gordon-mackerron/

Makhijani A 2013a *Traveling Wave Reactors: Sodium-cooled Gold at the End of a Nuclear Rainbow?* Institute for Energy and Environmental Research, Takoma Park, Ma http://ieer.org/resource/reports/traveling-wave-reactors-sodium-cooled-gold-at-the-end-of-a-nuclear-rainbow/

Makhijani A 2013b *Light Water Designs of Small Modular Reactors: Facts and Analysis', Light Water Designs of Small Modular Reactors: Facts and Analysis* (Tokama Park, MD: Institute for Energy and Environmental Research) http://ieer.org/wp/wp-content/uploads/2013/08/SmallModularReactors.RevisedSept2013.pdf

Muccusker P 2016 *Newcastle company at forefront of technology for small nuclear reactors*, Newcastle Chronicle, February 10th: http://www.chroniclelive.co.uk/business/business-news/newcastle-company-forefront-technology-small-10869984

NEA 2002 Accelerator-driven Systems (ADS) and Fast Reactors (FR) in Advanced Nuclear Fuel Cycles (Paris: Nuclear Energy Agency/OECD) http://www.oecd-nea.org/ndd/reports/2002/nea3109.html

NEA 2016 *Small Modular Reactors: Nuclear Energy Market Potential for Near-term Deployment*, Nuclear Energy Agency/OECD Paris https://www.oecd-nea.org/ndd/pubs/2016/7213-smrs.pdf

NNL 2010 *The Thorium Fuel Cycle*, UK National Nuclear Lab, Warrington http://www.nnl.co.uk/media/1050/nnl__1314092891_thorium_cycle_position_paper.pdf

NNL 2012 Small Modular Reactors: their potential role in the UK (Warrington: UK National Nuclear Lab) http://www.nnl.co.uk/media/1048/nnl__1341842723_small_modular_reactors__posit.pdf

Nuclear Energy Insider 2016 *US operator seeks swift SMR licensing to optimize low-carbon output*, Nuclear Energy Insider, April 29th http://analysis.nuclearenergyinsider.com/us-operator-seeks-swift-smr-licensing-optimize-low-carbon-output

NuScale 2016 NuScale compay's LCOE comparative estimates http://www.nuscalepower.com/images/nuscale_smr_benefits/Right_Column/nuscale-operating-costs.jpg

Phys.org (2012) *Thorium: Proliferation warnings on nuclear 'wonder-fuel'*, Phys.Org web news, December 5th http://phys.org/news/2012-12-thorium-proliferation-nuclear-wonder-fuel.html

Ramana M V and Mian Z 2014 *One size doesn't fit all: Social priorities and technical conflicts for small modular reactors*, Energy Research and Social Science, 2, pp115-124 https://www.academia.edu/8114310/One_size_doesn_t_fit_all_Social_priorities_and_technical_conflicts_for_small_modular_reactors and www.sciencedirect.com/science/article/pii/S2214629614000486

Roulstone T 2015 Presentation at the Small Modular Reactor Summit, London, 20-21 October http://www.nuclearenergyinsider.com/smr-uk/2015/conference-materials2015.php

Terrapower 2016 Company website http://terrapower.com/pages/technology

Thorium Energy World 2016b *India aims to build worlds first Thorium ADS*, Thorium Energy World news, July 19th http://www.thoriumenergyworld.com/news/india-aims-to-build-worlds-first-thorium-ads

Till C and Chang Y 2011 *Plentiful Energy -The story of the Integral Fast Reactor*, Create Space, extracts at https://bravenewclimate.com/?s=IFR

Toke D 2016 *Small Modular Reactors: wishful thinking on a grand scale*, Green Energy Blog, April 11th http://realfeed-intariffs.blogspot.co.uk/2016/04/small-modular-reactors-wishful-thinking.html

Tovey A 2016 *Rolls-Royce could power Britain's nuclear future with mini reactors*, The Daily Telegraph, March 19th http://www.telegraph.co.uk/business/2016/03/19/rolls-royce-could-power-britains-nuclear-future-with-mini-reacto/

Transatomic 2016 Company website http://www.transatomicpower.com/the-science/

UCS 2013 Small Modular Reactors: Safety, Security and Cost Concerns (Cambridge, MA: Union of Concerned Scientists) http://www.ucsusa.org/nuclear-power/nuclear-power-technology/small-modular-reactors#.WFcVV1d0Ndt

Weinberg 2012 Written evidence submitted by the Weinberg Foundation (NUC 20) to the House of Commons Energy and Climate Change Committee, July http://www.publications.parliament.uk/pa/cm201213/cmselect/cmenergy/117/117vw18.htm

Weinberg 2014 Weinberg Foundation, written evidence to the House of Lords Select Committee hearings on the Resilience of the Electricity system, REI0027, September https://www.parliament.uk/documents/lords-committees/science-technology/Resilienceofelectricityinfrasrtucture/Resilienceofelectricityinfrastructureevidence.pdf

Weinberg 2016 Weinberg Foundation website http://www.the-weinberg-foundation.org/learn/next-gen/msr/

Westinghouse 2016 SMR promotion on company website http://www.westinghousenuclear.com/New-Plants/Small-Modular-Reactor

WNA 2016 Small Nuclear Reactors (London: World Nuclear Association Information Library) http://www.world-nuclear.org/information-library/nuclear-fuel-cycle/nuclear-power-reactors/small-nuclear-power-reactors.aspx

WNN 2009 *Thorium-fuelled exports coming from India*, World Nuclear News, September 17th http://www.world-nuclear-news.org/NP_Thorium_exports_coming_from_India_1709091.html

WNN 2011 *Deep Sea Fishing*, World Nuclear News, January 20th http://www.world-nuclear-news.org/NN_Deep_sea_fission_2001111.html

WNN 2016a *Terrestrial Energy to complete US loan guarantee application*, World Nuclear News, September 14th http://www.world-nuclear-news.org/NN-Terrestrial-Energy-to-complete-US-loan-guarantee-application-1409167.html

WNN 2016b *UK considers how to use small reactor opportunity*, World Nuclear News, October 18th http://www.world-nuclear-news.org/NN-UK-considers-how-to-use-small-reactor-opportunity-1910161.html

WNN 2017a *NuScale makes history with SMR design application*, World Nuclear News, January 13th http://www.world-nuclear-news.org/NN-NuScale-makes-history-with-SMR-design-application-13011701.html

WNN 2017b *Terrestrial Energy unveils SMR licensing plans*, World Nuclear News, January 24th http://www.world-nuclear-news.org/NN-Terrestrial-Energy-unveils-SMR-licensing-plans-24011701.html

X-Energy 2017 *The Xe-100 will fundamentally shift where—and how—we use nuclear energy*, X-Energy Company website http://www.x-energy.com

IOP Concise Physics

Nuclear Power
Past, present and future
David Elliott

Chapter 4

Nuclear power revisited

We face some major energy policy choices, and a need to move away from fossil fuels, with nuclear energy and renewables often being presented as supply-side solutions, but also as polar opposites. Assuming energy demand can be managed appropriately, it seems likely that, in time, renewables can supply most energy needs at reasonably low cost. So why is there still support for nuclear power? The simple answer is that it is well established with powerful supporters, and it does deliver energy, about 11.5% of global electricity, with relatively low carbon emissions and relative reliability. For some, the vision of a nuclear future remains strong and new ideas are being promoted, even if many of them in fact are recycled old ideas.

As we have seen, there were often good reasons why some of the old ideas were not developed further at the time. While that does not mean that none can be, any candidate new nuclear option has to face the economic, safety and security problems that have slowed the progress of nuclear so far. Nuclear power certainly can deliver relatively low carbon energy, but it does have some major disadvantages. They include safety and security issues, as well as economic problems. As we have seen, some of the new candidate options discussed above seek to deal with or limit some of these problems. How well do they do that?

4.1 A review of the prospects for new nuclear

Most new technologies of whatever sort are often claimed by proponents to be cheaper than what went before. Nuclear technologies are no different. The starting position is one where costs for the existing range of reactors seem to have escalated continually, with, in effect, negative learning curves, at least in the US and France (Boccard 2014). That may not always be the case everywhere, as was claimed recently (Lovering *et al* 2016). Though that study has been challenged as unrepresentative and flawed (Gilbert *et al* 2016, Koomey *et al* 2016).

It may be that the new plants can be built more cheaply in Asia, where labour costs are usually lower and, arguably, safety regulation less onerous, but we will have to wait and see what becomes of the Generation III projects there and the few new ones in Europe and the US. Will they avoid the problems of Generation II? Some critics have technological doubts e.g. on the AP1000 (Roche 2016). Questions have also been raised about some Generation III passive safety features (IRNS 2016). Even if they can overcome the technical problems, will they, and the new Generation IV projects, escape what seems to be the general historical trend towards increased prices? The EPR clearly is not doing well, but that certainly is the aim of nearly all the newly proposed Generation IV technologies: they hope to break the mould technically and economically. Since none yet exist, it is hard to say if they really can achieve this, and perhaps more importantly, if so, on what timescale. Past experience with small reactors and the limited number of thorium and fast reactor projects may not be a good guide: as we have seen, many of them failed technically or economically, but that may not mean that new attempts will necessarily share the same fate. That may depend on the resources allocated to this work.

It is interesting in that regard that, in the conclusion of a review of the practical problems of molten salt thorium reactors, French physicist Hubert Flocard says: *'In my opinion, thorium and molten salt reactors technologies belong definitely in the domain of research. They certainly have a potential, which deserves scientific and technical investigation. On the other hand, given the present situation of the nuclear energy research institutes of the western world and the general decline in their enrolment of high-quality well-trained young engineers, it is improbable that much work will be invested into such an innovative, far-reaching but also risky options. Therefore, if nuclear energy is to provide a significant contribution to the world energy mix of the 21st century, it is doubtful that thorium and molten salt technologies will be ready in time to take part'* (Flocard 2015).

One of the key issues that would need attention is safety. Arguably, that is clear from the history of nuclear power so far. Some in the industry complain that the safety standards required are excessive and unnecessarily tight compared with those for similar industries, but many people feel that nuclear risks are unique and must be reduced as far as possible.

As we have seen, the Generation IV International Forum has looked at Generation IV options, and designs that might be developed *'by the second half of this century'*. It chose six candidates: sodium-cooled fast reactors (SFR); very high temperature reactors, with thermal neutron spectrum (VHTR); gas-cooled fast reactors (GFR); lead-cooled fast reactors or lead-bismuth eutectic (LBE) cooled fast reactors (LFR); molten salt reactors (MSR), with fast or thermal neutron spectrum; super critical water reactors (SCWR), with fast or thermal neutron spectrum. A review by the French agency IRSN said that it was hard to see how any full scale MSR (or SCWR) could be built *'before the end of the century'*, while, although more work was still needed, the SFR system was the only one to have reached a degree of maturity *'compatible with the construction of a Generation IV reactor prototype during the first half of the 21st century'* (IRSN 2015). That may be true (as noted, there have been several prototypes), but it has to be said that not everyone would agree that liquid sodium-cooled fast reactors are the safest option. Indeed, the IRSN said that, at the present stage of development, it did not see any evidence that *'the systems under review are likely to offer a significantly improved level of safety compared with Generation III reactors, except perhaps for the VHTR'*, although even that would require *'significantly limiting unit power'*.

An issue linked to safety is the management of radioactive wastes and spent fuel. Arguably, the industry has not done too well in this area so far. No site for the permanent disposal of high-level long-lived wastes yet exists anywhere in the world. Those that have been proposed are very expensive—tens of billions each on current estimates. It is true that the waste that has already been created has to be dealt with, but arguably it would be wise to avoid producing more, given that it is unclear if any of it can be secured for the vast time scales necessary to ensure it offers no risks to future generations (Blowers 2016, NDA 2016).

In the case of the UK, a site for final geological disposal of its high level nuclear waste is still being sought, with communities being invited to host it, possibly in return for substantial funding for local social projects. However, it would be earmarked preferentially for the existing/current legacy waste, possibly to be loaded up from around 2060 onwards. There would not be room for the wastes from the new Generation III plants, currently proposed to start up in the late 2020s, until around 2130. So the wastes from the new plants will have to stay in interim stores somewhere, probably on-site, for a long time after the new plants have closed, even assuming 60-year plant operating lives (HMG 2013).

As we have seen, some of the Generation IV candidates are claimed to be able to reduce the waste burden. For example, it is claimed that, with fast breeders, some of the wastes they produce can be burnt up or pyro-processed, but, although looked at in the US, the latter has not been widely tested or used (Cochran *et al* 2010) and the idea of using *Prism* type fast reactors for plutonium/waste burn up has not so far proved popular (Green 2014). With thorium molten salt reactors, it is claimed that waste handling is made easier, since, whereas in solid-fuelled reactors, fission products accumulate within the fuel rods, in MSRs, they can be continuously removed, and there may also be fewer of them, though they still have to be dealt with.

Waste disposal also touches on the wider issue of environmental impacts. One of the key benefits claimed for nuclear power is that nuclear plants do not generate carbon dioxide gas. While that is true, the mining and processing of their fuel is very energy intensive, and for the moment most of this energy comes from fossil fuel. The result is that there are associated carbon emissions, which will increase as high grade uranium ore becomes more scarce and resort has to be made to less accessible and lower grade ores, requiring progressively more and more energy for mining and processing.

Using non-fossil fuels, renewables or of course nuclear power itself, for this energy would avoid these emissions, but even so, there will be diminishing returns. At some stage, we reach the so-called *point of futility*, when more energy is used in production of the fuel than is generated by using it in reactors (WISE 2015).

At present the energy return on energy invested (EROEI) ratio for the nuclear fuel cycle based on PWRs is around 15:1. It has been suggested that it would fall to 5:1 with lower grade ores and possibly lower, depending on the extraction and processing technology used. Modern *in situ* ore leaching techniques and laser enrichment systems use less energy. However, for comparison, renewables like wind turbines do not need any fuel to run and their overall EROEIs are generally much higher, for example they have been put at up to 80:1 for modern wind projects at good sites and improving as the technology develops (Harvey 2010).

There are inevitably some disagreements about EROEI estimates: they do depend on the boundary condition assumptions. For example, some analysts ignore the input operational fuel energy and just focus on embedded energy in the hardware and construction materials. That inevitably privileges nuclear over renewables. The discussion on methodology continues (Weißbach *et al* 2013, Raugei 2013, Weißbach *et al* 2014, Raugei *et al* 2015).

The use of the more abundant thorium resource might improve the EROEI for nuclear for a while, though, since it has to be converted to U233, it is less energy productive, depending also on reactor type. One study suggested that fuel production related carbon emissions for a thorium-uranium-fuelled EPR would be around 4%–7% higher than from a uranium-fuelled EPR. Moreover, emissions from a Th-U-fuelled heavy water plant and a Th-U-fuelled modular helium plant were typically around 20%–100% higher. It was also noted that while some thorium resources may be easier to mine, and so have a lower energy/carbon debt, the scale of these resources was limited (Ashley *et al* 2015).

Nuclear fuel reserves of whatever sort are of course finite. Indeed, without breeders, there may only be sufficient uranium for a few decades-worth of supply, if the use of nuclear power expands. Moreover, breeders, whether using fertile uranium 238 or thorium, still need fissile fuel (or accelerators) to provide neutrons initially. By contrast, renewables are not resource limited, so in the long term there is no contest.

In the short to medium term, some say we should and can use both nuclear and renewables. However, there are inherent conflicts, and not just in terms of competition for funding. Most types of nuclear plant cannot vary their output rapidly or regularly without safety and cost problems. In addition to heat stress issues with regular ramping up and down, it takes time for the activated xenon gas that is produced, when reaction levels are changed, to dissipate; it can interfere with proper/safe reactor performance. In any case, the current types of large nuclear plants need to be run 24/7/365 to recoup their large capital costs. They can just about deal with some of the regular daily energy demand cycles (demand peaks in the evening, low demand at night), but not with the irregular fast variations likely with large amounts of wind and solar generation on the grid—they cannot be used to back up the short-term variable output from these renewables. It is conceivable that they could cover the occasional longer periods when wind and solar is at minimum, but that would mean running at lower levels at other times, ready to ramp up slowly to meet the lull periods, undermining their economics.

If nuclear plants cannot vary their output significantly, then there will be a direct conflict as renewables expand. They will not fit well together on the grid. If there is a large nuclear input and also a large renewables input, there will be head to head operational conflicts when energy demand is low e.g. at night in summer, when, in the UK, demand is around 20 GW. The UK is aiming for 16 GW of new nuclear in the 2020s and more later, taking it well beyond 20 GW, and also 30 GW of renewables by around 2020, more later. So, assuming the excess cannot be exported, or stored, which will be turned off when demand is low?

Can any of the new nuclear options avoid these problems? It is claimed that small modular reactors can vary their output more easily, but there may still be an economic penalty for this. However, if they also supply heat as well as power, in CHP mode, that may improve their flexibility, but, as argued above, there may be limits on sites near heat loads.

It is sometimes argued that molten salt reactors will be more flexible, being able to adjust activity levels easily without major stresses, given that operations are done within a fluid matrix. That has yet to be proved, as has the economic viability of this concept. Some early claims about potential performance have certainly proved to be wrong (Temple 2017). While that remains unresolved, it could be argued that, in general, there are many cheaper and easier ways to balance variable renewables on both the supply and demand side (Elliott 2016).

It is possible though that some of these balancing measures may also help deal with inflexible nuclear. For example, in addition to export of surpluses, storage is an option. It is hard to store electricity in bulk, but, as with renewables, surpluses could be used to charge electric vehicle batteries at night, with power being exported from them back to the grid, if needed. Some nuclear projects may also find a role in producing hydrogen, which is easier to store, and that could help with balancing variable renewables. For example, the surplus output from nuclear plants could be converted into hydrogen gas, by electrolysis of water, and then perhaps converted into methane gas or other synfuels, using CO_2 captured from the air or power station exhausts. Germany is already doing this with several wind-to-gas/power-to-gas plants, some feeding methane gas into the national gas main.

It has been argued that if there is a large already-built and paid for nuclear component (as in France) it could be used to balance some of the variations in renewable availability (Cany *et al* 2015). That seems to be just a way to sustain the over-large nuclear fleet for a bit longer. It would arguably not be economic to build *new* nuclear plants to do this directly with electricity, but it might be economic to build some to make hydrogen by electrolysis for grid balancing, and/or for vehicle fuel and other synfuel production, feeding into the so-called hydrogen economy (ACS 2012). However, they would be in competition with renewable electricity generators, including those (like hydro and geothermal) that can provide continuous grid input and those that are variable, but at times producing surplus input.

As an alternative, it is possible that high temperature reactors could produce hydrogen by high temperature dissociation of water, although they would be in competition with focused-solar plants that can also do this in some locations. However, nuclear plants would have the advantage that they could be located in places where solar irradiation was weak.

Some nuclear plants could also be used for reformation of methane gas to make hydrogen. For example, in relation to an ambitious 'high nuclear' scenario published on the *Energy Matters* website, it is suggested that *'molten salt reactors will produce heat at a high enough temperature for steam methane reforming. The resultant CO2 could then be stored, and the hydrogen used to supply energy in the winter (either for direct heating, or by electricity generation in gas turbines or fuel cells)'*. Moreover, it is suggested that some high temperature reactors, such as the pebble bed design being built in China, *'may be able to switch from the production of electricity to the thermal production of hydrogen from water, whenever demand is low'*. However, it is also noted that in both cases, *'spare capacity would be in the summer, and demand for hydrogen will be highest in winter, and storing large quantities of hydrogen is not straight forward'*. Then again it is pointed out that *'high temperature reactors, including molten salt reactors, can store energy in the form of hot salts'* and it is noted that the developers of the Moltex Reactor have proposed storing *'several hours' worth of thermal output'*, with the plants varying their electrical output in the meantime (Terrell and Dawson 2016).

Clearly there are many interesting possibilities. Although using smaller plants in a variable output mode has attractions in terms of grid balancing, it also seems likely that hydrogen production would be best done with large scale units e.g. in large dedicated IFR-type plants. They could provide hydrogen as a fuel for vehicles and/or as feedstock for industrial processes. That would capitalise on the fact that nuclear plants are best used to provide continuous output and can be large, and, in theory, located near industrial demand centres. Even so, there might also be merit in having mixed, and variable, power, heat and hydrogen outputs, depending on location. In the case of heat production, it would clearly be vital to be near large heat loads, whereas in theory hydrogen, produced in bulk, but variably, could be piped or shipped long distances and electricity can also be transmitted over long distances.

However, in some of these cases, once again, there could be direct competition from renewables. Some renewables (solar, geothermal and biomass) can supply heat direct to users. Some companies are using large on-site wind turbines for direct power. Roof-top PV solar arrays are being widely used for the power needs of factories, commercial/retail warehousing and parts centres. Medium scale renewables, like multi-MW wind, tidal and solar farms, feeding power to the grid, alongside smaller local projects, can supply some industrial complexes, even if they are located far off, as in the case of the new large 1 GW concentrated solar thermal plants that are being built in desert areas: long distance HVDC supergrid energy losses are relatively low. Nevertheless, there are advantages in having a power supply nearby, for large loads like steel mills or aluminium smelters, and also for supplying heat, and nuclear plants might be suited to that in some locations (WNA 2016).

Looking further into the future, some of the same options might be open to fusion plants, if they prove to be viable, as in the 'fusion island' industrial-energy complex idea, with very large fusion plants perhaps supplying hydrogen for shipment in bulk (Nuttall 2005). On a more local basis, some also look to mini-fusion plants (Tyler 2016).

All of that is very speculative. For the moment however, moving back to fission and current reality, as we have seen, nuclear power is faced with a range of pressing problems, including the need to deal with long lived wastes, as well as the cost of maintaining safety and security, problems which are either absent or relatively small for most renewables, the cost of which is falling as markets build and they move down their learning curves. Most renewables are well ahead in this regard, despite not having had the high R&D support enjoyed by nuclear, but it is possible that new nuclear technology will at some point move down its learning curve to lower prices, and also resolve some of the problems currently facing nuclear.

While some of the current range of Generation IV options may thus thrive and be competitive in some contexts, it is far from clear whether the fortunes of nuclear power can be revived substantially, unless it breaks into new types of energy market, as outlined above. Even then it may be limited to niche applications, some of which may be contested by emerging renewables. So its longer-term future remains open to debate.

4.2 What long-term future for nuclear?

One current view is that nuclear power will continue to play a role into the future, although its growth may be less than was at one time expected, with renewables playing much more of a role. Thus the World Energy Council's 2016 *Grand Transition* report outlines a scenario in which solar and wind supply 39% of global power by 2060, and nuclear 17%, in a context where electricity demand doubles by 2060, although primary energy use stabilises before 2030 (WEC 2016). For comparison, in its '2 degrees' scenario, the International Energy Association has nuclear at 17% by 2050 with wind and solar supplying 30% of global electricity, although it adds that 'significantly more' could be done in some countries (IEA 2016). Moreover, with hydro and other renewables added in, the total would be much higher.

Even so, these projections for renewables are much lower than those from many academics and NGOs, and are also much higher for nuclear. There are now several global and national scenarios with renewables supplying nearly 100% of electricity by around 2050 and nuclear all but phased out (Wolff and Jones 2016). Some countries are aiming for that (Payton 2016). By contrast, some nuclear enthusiasts look to a world dominated in electricity terms by nuclear, with fast breeders used to extend fissile fuel reserves, and then possibly fusion adopted worldwide. So there are competing visions of the future (Elliott 2015).

It is conceivable that technological advance could solve some of the problems with Generation IV designs identified above at some point in the future. It is also possible that fusion might then take over longer term. Fusion might be seen as Generation V (see box 2.3). Indeed, some optimistically include it in Generation IV, and there are several small companies and groups working on novel ideas, as well as the US laser-fired pellet compression system. It is something of a race, though on a long timescale, 60 years and counting so far, with the more conventional ITER in France being the next big step (Carrington 2016). Unless someone else gets there first (Lockheed Martin 2016). Certainly some think the whole long running and very expensive fusion R&D process could be redirected along different and maybe faster routes than the current ITER 'tokomak' magnetic containment approach, which does have its problems (Hirsch 2015, 2016).

Fusion reactions do not generate fission products, but the large neutron flux that is generated in deuterium–tritium reactions will activate reactor components and containment materials, which will have to be periodically stripped out to avoid interference with the fusion process. So there will still be some active wastes to deal with. The fact that fusion reactions are hard to sustain does in principle mean that if there are containment or other operational failures, the reaction will shut down instantly, but there may still be safety risks e.g. potential for hazards from the accidental release of radioactive tritium gas. The issues will obviously be different for some designs, but in all of them, getting useful energy out will be a problem: is the idea just to boil water again, as with fission (and coal!) plants?

As for fuel resources, while there is plenty of deuterium in the sea, tritium, made from lithium, may be harder to secure longer-term: lithium is in increasing demand for electric vehicle batteries. More exotic materials may be available off-planet, but here we are moving into science fiction territory. It could be asked, why go to the asteroid belt to get helium 4 or whatever, when we already have a fusion reactor in the sky, supplying the Earth with more free solar energy that could ever be needed? It is possible that use will be found for nuclear fusion and perhaps advanced fission, if human settlements are set up on other planets, particularly on those a long way from the Sun. On Earth itself, it is less clear, unless there are major technical breakthroughs.

Absent that, electricity generation from renewables seems likely to dominate long-term, balanced by long-distance HVDC supergrid transmission links and hydrogen production and storage, using the occasional surplus outputs from the variable renewables. Renewables can also meet most heat loads, either directly or via electricity; similarly for transport, via electricity, liquid biofuels, biogas and green syngases like hydrogen, although that may be harder. Some environmentalists do not want too much use to be made of biomass, given potential conflicts with food production, biodiversity and carbon balance issues. However, it is interesting that one set of 2050 scenarios, produced by researchers at Stanford University, suggests that, given proper attention to energy saving, near 100% of global energy, including heat and transport as well as electricity needs, could be met from wind, water and solar energy, without the use of biomass energy (Delucchi and Jacobson 2013).

Scenarios like this are about what *could* be done, but we also need to think in terms of what *will* be done. Progress with renewables has been increasingly rapid, driven by falling prices, with, in the best sites, wind and solar now generating at around 2–3 US cents kWh^{-1}. Average prices will obviously be higher, but wind power cost have fallen by 50% since 2009 and PV module cost by 80% since 2008 (Leibreich 2016).

However, while electricity use is falling in the West (e.g. by 13% in the last decade in the UK) and energy efficiency measures can and should reduce it further, demand for energy overall is increasing in some parts of the world, with, for example, industrial and commercial growth in Asia being partly driven by the use of fossil fuels, coal especially, but also oil products in vehicles. That is leading to major and very visible air pollution and major health problems, and will, in time, add to the climate change problems we all face. The big strategic issue is whether the newly rising economies of the East can make the transition to renewables, and efficient energy use, fast enough to limit or avoid these problems. Many are trying: they have an obvious and urgent air pollution incentive, and as costs continue to fall, the transition to renewables also begins to make even more economic sense, both in terms of national clean energy use and also, crucially, technology export potential. China has clearly understood that in terms of PV solar—it is exporting PV around the world.

Renewables do seem to be seen to be the main way ahead for many countries. However, in terms of attempting to deal with climate change, there are some rival energy supply options, apart from nuclear. The development of carbon capture and storage (CCS) clean-up technology has been much touted as an interim measure, allowing for the continued use of fossil fuels, but it is not yet widely or even marginally adopted. It is expensive, with as yet unresolved issues, e.g. can carbon dioxide gas be stored in large volumes in underground strata safely and indefinitely? Given that fossil fuel reserves are finite and geological storage sites are also limited, CCS does not seem to offer a long-term energy solution. Using biomass feed stocks for combustion in power plants with CCS, does offer the intriguing possibility of carbon *negative* energy production, but there would still be limits to that, in terms of CO_2 storage space, and there are land-use and biodiversity constraints on biomass growing.

There are also cost and environmental constraints on some of the more extreme technical fixes for climate change, for example involving reducing incident solar radiation by injecting aerosols or reflective particles into the upper atmosphere or even putting giant solar shades in geostationary orbit. It should be obvious that, rather than trying to block out the Sun, so that we can continue to burn fossil fuels, we can and should use the energy it provides.

For the moment then, the main low-carbon energy supply options are renewables and nuclear. While some Western countries are now aiming to get 80% or more of their electricity from renewables by 2050, and the global picture is also promising, nuclear power may still find a role in electricity production, as is already clearly the case in Asia. How fast and how far it can expand there and elsewhere is unclear, given its problems, the rise of local opposition, and the generally much more rapid expansion of renewables in the electricity market. While pursuing renewable growth, China, India and South Korea may also continue to expand their use of nuclear, but a 2015 IEA report noted that nuclear power would only account for around 1% of electricity generation in south-east Asia (the ASEAN area) by 2040 (IEA 2015). Moreover, Vietnam has now abandoned its plan to build a Russian-backed plant due to its high cost, while WWF says that Vietnam could get 80%–100% of its power from renewables by 2050 (WWF 2016); so could India and China (WWF 2013, 2014).

So longer-term, the nuclear industry may have to look to other areas and other markets. It is interesting that some members of the nuclear fraternity have concluded that nuclear plants of whatever type are probably not going to be competitive with renewable electricity in the long term, or even quite soon. If nuclear is to have a longer-term future it may have to find new roles and new markets, including in areas and uses where renewables may not be so effective or sufficient. For example, as noted earlier, some nuclear plants could find a role in hydrogen production for vehicle fuel or as a feedstock for industrial processes. That seems to be part of the approach the US has taken in recent years, in terms of nuclear R&D effort (MIT 2009).

It would make sense, technically and economically, to run nuclear plants flat out to do this, rather than varying their output: and in that mode they would not be much use as backup for variable renewables. But any hydrogen they produced might help for balancing renewables. So they could helpfully coexist. As we have seen, heat production is another option, although that might be more locationally constrained, except perhaps for use for industrial processes.

To summarise, while renewables may prove to be effective in at least some of these new roles, some of the Generation IV design may also be suited, although, in both cases, and in terms of the wider energy context, there is a way to go to move from concepts to reality. There were bold ideas in relation to nuclear in the 1950s, but as we have seen, little came of them. It is of course also possible that some of the bold ideas now being touted in relation to renewables and their integration will also fail. Long distance supergrids may prove to be uneconomic, smart grid balancing similarly. It is certainly all very technically challenging. Box 4.1 summarises the current state of play in terms of nuclear versus renewables, and the issues raised are explored in more detail in the appendix.

Box 4.1. Nuclear and renewables—a summary comparison.

Resource base—uranium and thorium reserves are finite, solar and lunar energy is not. But breeders may extend the viability of the fissile fuel resource and fusion may be less resource limited.

Economics—renewables are mostly becoming cheaper than nuclear, depending on the context, but new technologies and breakthroughs in both cases may change the situation. So could cutting demand.

Safety and risks—major nuclear accidents can have large social, economic and ecological impacts, while the risks and impacts with renewables (hydro apart) are mostly low and local.

Security—nuclear plants and materials are prime terrorist targets, requiring massive security. That can hardly be said of most renewables, although large hydro dams are sometimes a military target.

Markets—nuclear may not be competitive in the emerging electricity market, as renewables expand, but might be able to compete in the heat and synfuel market, although that is unclear.

4.3 Conclusions: the way ahead

The choices of the route ahead depend on many factors. As box 4.1 illustrates, economics is only one. There are also divergent views about safety and environmental impact. It is clear that the combustion of fossil fuels, coal especially, creates major social and environmental problems due to air pollution and climate change. Nuclear advocates say a switch to nuclear power can avoid these (Kharecha and Hansen 2013). But so can a switch to renewables, and arguably with fewer downsides, for example in terms of health and safety risks. People do fall off ladders when putting solar collectors on roof-tops or get injured or even killed when installing or maintaining wind turbines, but so far the global total for occupational deaths from working on wind projects is around 150, with no recorded injuries to the general public from wind farm operation or accidents. The same cannot be said for nuclear plants.

The full analysis of social and environmental impacts clearly has to go beyond the anecdotal, and review such issues as visual intrusion, land-use and wild life damage. However, while large catchment areas may be needed (MacKay 2009), working with low density energy flows, means that impacts in general seem to be relatively small and manageable (Clarke 2016), though some are less so: hydro dam failures can be disastrous and some biomass production and use can have significant ecological impacts (McCombie and Jefferson 2016).

While the relative scales and significance of these impacts can be debated (Elliott 2013), it is clear that nuclear disasters can have major, wide ranging, impacts, and it seems likely that if, tragically, there is another major nuclear accident somewhere in the world, or a major terrorist-related nuclear incident, the prospects for nuclear developments of whatever sort may decline even further. Technical, deployment or operational problems will no doubt emerge with some renewables, but, given that there are so many different technologies, with few common features other than using natural energy flows, it is hard to see how generic impact problems could undermine the overall development of renewables.

Debates over issues like this will no doubt continue, with, on the nuclear side, some of the disagreements reflecting differing views on the risks of radiation exposure, based on differing models of short and long term biological impacts, and the role of continued internalised exposure from radioactive materials that are ingested, breathed in or absorbed, compared to one-off external radiation exposure episodes (CERRIE 2004). Location is also important. There can be cumulative impacts from low-level emissions near plants. Moreover, while, despite assertions of low risks, almost all nuclear plants have (so far) been sited well away from cities, that sort of protection is not available to those who work in, or live near, uranium mines, with their radioactive tailings (Rosen 2016).

On the renewables side, there may be differing environmental perceptions and social values: some see wind farms as glorious symbols of environmental sustainability, others see them as ugly industrial blots on the landscape. While it is clear that some people find the look of wind farms offensive, it is hard to imagine many seeing nuclear plants as aesthetically pleasing.

However, the issues are much larger than visual intrusion, other local impacts including land-use concerns, or even the more prosaic economic assessments. At base it concerns our faith in the reliability of the technology and its ability to deliver energy safely and effectively into the future. In the end, the choice of the way ahead may come down to technological confidence and perceptions of competence, and associated perceptions of risk. Building complex nuclear devices involves trying to push technology, pressures, temperatures and materials to the limit. Working with molten metals and high radiation fluxes certainly taxes our expertise, and trust. Though, in different ways, so too does developing smart interactive grid-management systems based on variable, diffuse renewable sources. See the appendix for a comparison.

The two approaches also reflect different views of the future. It is conceivable that nuclear power could support a more decentralised community-orientated society, but renewables lend themselves much more to this, PV solar especially, and seem likely to raise fewer safety and security issues. They also do not create radioactive wastes. While some argue that renewables will be less productive in energy terms and so will require a more frugal, and perhaps even coercive, society, others see nuclear technology, with its security risks, as leading to a more centralised and authoritarian society. An interesting social and political debate (Elliott 2015).

The economic debate will also continue, although these days, with renewable costs falling and capacity auction processes becoming common, nuclear projects will be hard pressed to compete without continued public subsidies. Some have looked to carbon taxes and the like rescuing nuclear power, but the Nuclear Economics Consulting Group says: *'Politically feasible carbon pricing is not likely to provide the long-term revenue needed to support existing or new nuclear power projects. Carbon prices are likely to be too indirect, too late, too low, and too uncertain to provide real financial support for nuclear power projects'*. Instead, it said, more direct targeted support is needed, since, by itself, *'the market fails to provide support for nuclear'* (WNN 2016a).

Loan guarantees have been provided in the USA to help new projects get going, but that has not been too successful—higher levels of investment support was needed. Nuclear industry lobbyist Steve Kidd has commented: *'Even with government incentives such as loan guarantees, fixed electricity prices and certain power off-take, nuclear projects today struggle to make economic sense, at least in the developed world'* (Kidd 2015).

In recent years, loan guarantees apart, most Western governments have been unwilling to provide direct financial support from taxes for nuclear projects, as some did in the early days, now preferring indirect forms of support via levies on consumers, as with the UK's CfD subsidy system, which is being offered as a way to support Generation III projects. We have yet to see if that will be successful, and how it, or some other form of support, might be applied to Generation IV projects, when and if they emerge from the R&D phase. A US consortium has called for public-private finance schemes to support SMRs (WNN 2017).

In the past there have been bold plans for global programmes of nuclear roll out, as in the 1950s *Atoms for Peace* initiative backed by the US. More recently, in the

2000s, President George W Bush promoted a Global Nuclear Energy Partnership (GNEP) under which the US (and maybe others) would help to service civil nuclear developments globally, including possibly with fuel supply and waste processing.

In one variant of this approach, the US or other vendors might install reactors, possibly small modular reactors, in developing countries, to be run on a franchise basis, the reactor modules being leased, and the fuel/waste being controlled, by the vendor. For example, the US might repatriate the spent fuel capsules and reprocess the spent fuel to extract plutonium for use in its own reactors. Certainly at one time there was talk of SMRs being rolled out across the developing world, with micro projects being seen as well suited to countries without well-developed grids. Echoing the rhetoric of the *Atoms for Peace* initiative, at one point US Energy Secretary Samuel Bodman claimed that *'GNEP brings the promise of virtually limitless energy to emerging economies around the globe'* (DoE 2006).

However, given the changed security climate after the 9/11 terrorists attacks on the US, and increased concerns about the risks of proliferation of nuclear weapons material, GNEP seems subsequently to have been sidelined. The GNEP programme did not prosper under President Obama. Instead there has been an expansion of more conventional commercial approaches. For example, France, Russia, Japan, China and South Korea are seeking to export their nuclear technology around the world.

As we have seen, what they have on offer varies, as does their success in winning orders. With markets for new nuclear somewhat limited in much (but not all) of the EU and in the US, the emphasis has been on Asia and, to a lesser extent, Africa, Latin America and the Middle East. It remains to be seen if these initiatives will bear fruit, and underpin a Nuclear Renaissance, or whether the renewable options will dominate. It certainly might seem odd for countries and regions blessed with high solar potential to invest in nuclear, unless there were other attractions than just energy. The potential civil–military nuclear links are hard to avoid.

While strategic issues like that have still to be resolved, in terms of the technologies and their future prospects, looking broadly, we seem to be faced with a choice between two visions of the future. Both have their problems. The 'high nuclear' vision may be unworkable, as may the 'high renewables' vision. However, we have been here before, with early visions of nuclear cornucopia. The intervening period may have taught us to be more careful about overly optimistic views, but it remains the case that optimism is important. It provides motivation, while pessimism can be stultifying. Even so, we do have to agree on the direction of travel, and on which vision to aim for.

Nuclear power has had a long and somewhat turbulent history in the UK, but it is currently favoured by the government, although the economic and political uncertainties remain: the financing of the Hinkley project has been particularly controversial, with the outcome still uncertain (Taylor 2016). Globally, the policy battles go on, with some nuclear lobbyists arguing that it is vital to back nuclear expansion to deal with climate change, in the extreme claiming that renewables cannot do that (EP 2016, Weinberg 2016, Neutron Bytes 2016).

There is certainly no shortage of promotion of nuclear expansion, with some exponents clearly still projecting confident views (WNN 2016b). In direct opposition, some green energy lobbyists argue that nuclear is a distraction, which cannot help much to limit climate change (Squassoni 2017), whereas, despite claims to the contrary, renewables can (Lovins 2015, Diesendorf 2016). I have to say that, on balance, I share the latter view: despite the constraints and challenges, renewables are developing well and look like being the way ahead for most if not all countries (Elliott 2013).

However, there is still a debate on the role of renewables and also on the potential of nuclear power, for example in terms of responding to climate change (Mecklin 2017). Hopefully this short study will have helped clarify the choices and issues a little. The story it tells is of repeated attempts over many years to develop a safe, economically and environmentally viable energy system based on using nuclear energy to generate electricity. As we have seen, despite massive funding, the results so far have, arguably, been mixed.

Certainly nuclear power has sometimes been attacked for not having lived up to the perhaps unrealistic expectations raised in the early days, with for example, US Atomic Energy Commission Chairman Lewis Strauss, in a 1954 address to science writers, claiming that *'it is not too much to expect that our children will enjoy in their homes electrical energy too cheap to meter'* (Strauss 1954). He was not specific about whether that was via fission or fusion (Wellock 2016), but it is clear that, either way, he has been proven wrong: it has not got cheap. Indeed, it seems to be getting more expensive.

In terms of technology, the substantial and long running efforts of many highly qualified and motivated people notwithstanding, while there have been some notable successes and reactor systems which have gone on to provide many years of relatively reliable service, there have also been some major disasters with commercial projects and repeated technological failures as new ideas were explored. One result has been a loss of faith in nuclear technology amongst increasing numbers of people. As might be expected, that includes most environmentalists.

Their objections often go beyond just concerns about environmental impacts and safety, important though they are: the basic technological trajectory is seen as flawed, with nuclear power being viewed as an unreliable way to respond to climate change. The environmentalist criticism also extends to the more specific technological development issues covered in this text. For example, leading UK environmentalist, Jonathon Porritt, has commented that *'the consistent history of innovation in the nuclear industry is one of periodic spasms of enthusiasm for putative breakthrough technologies, leading to the commitment of untold billions of investment dollars, followed by a slow, unfolding story of disappointment caused by intractable design and cost issues. Purely from an innovation perspective, it's hard to imagine a sorrier, costlier and more self-indulgent story of serial failure.'* (Porritt 2015)

That may be put a little aggressively, but a widely shared view is that the nuclear lobby is forever offering *'jam tomorrow, but never today'*, as the late Lord (Walter) Marshall, one-time head of the UK Atomic Energy Authority, once wryly admitted in the context of hoped-for cost reductions, adding: *'The British Public have never had the cheap electricity that we have always promised from nuclear power'* (Marshall 1987). That seems to have been a general pattern: the next reactor system will be better, cheaper, safer! After 70 years or so of development, regular assertions like that begin to ring a little hollow. It remains to be seen whether the next iteration will do any better. On past performance, it is hard to be optimistic.

Some countries will continue to press ahead with, and rely on, nuclear power, as a minority (30 countries out of the 196 global total) do at present, but it is likely that most will not want to go down this route and more that have already done so will back off. Some would see that as tragedy, given the huge effort that has been put into nuclear power development over the years. It would certainly be painful to admit failure. But the contribution that has been made to physics and also engineering will remain valid, and there would be some useful spin off, both technologically (there is some expertise overlap with other energy options) and policy-wise. For example, the long tortuous history of nuclear power may provide a useful warning for those of us that hope and believe that renewables can do much better. Technological innovation is not easy and we can expect similar up and downs, though hopefully with fewer major disasters and more technical and economic success. There is a way to go, but, along with energy saving, that does seem likely to be a more productive route forward.

Afterword: insider views

The analysis presented in this book suggests that nuclear power may have a limited role in future. That view is buttressed by recent developments in the nuclear sector, with several countries, led by Germany, abandoning nuclear programmes, and major power engineering companies like Siemens, RWE and E.ON in Germany, Engie (formerly Gulf Suez) in France and SSE in the UK, moving away from nuclear and focusing instead on renewables. Companies that have stayed with nuclear, such as EDF and Toshiba, are facing significant economic problems, with major new projects like EDF's Hinkley EPR becoming increasingly uncertain financially. In 2016 the UK's *Economist* magazine commented: *'Britain should cancel its nuclear white elephant and spend the billions on making renewables work'* (Economist 2016). They were not alone in seeing the cost as being too high; see figure 4.1.

More generally, in addition to the worsening investment climate for nuclear projects, public opinion polls suggest that, globally, most people do not support nuclear power, and even in the case of the UK, only 33% supported nuclear in a recent poll (BEIS 2016a), while, in another recent poll, only 25% backed the proposed Hinkley project (Populus 2016).

However, unsurprisingly, these views are not likely to be shared by many scientists and engineers in the nuclear field, or possibly by other scientists and engineers. Certainly many remain convinced that nuclear power is vital for the future, with the threat of climate change adding an extra impetus. In a Vision Prize poll in 2014, open to 'climate scientists and other scientists or researchers with relevant expertise' and backed by two US universities and the UK Institute of Physics, a strongly pro-nuclear view emerged. Of the nearly 100 scientists who participated, 71% agreed with the view, put in an open letter by four leading US

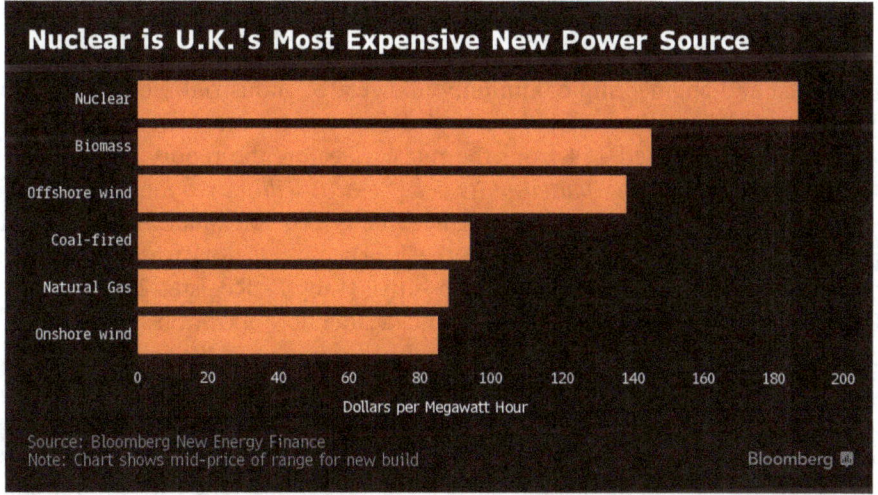

Figure 4.1. Bloomberg New Energy Finance data for UK new build Generation costs, mid price range (Morison and Shankleman 2016).

scientists (Emanuel 2013), that nuclear power was a critical component of any realistic plan to achieve climate stabilisation (ERW 2014).

That is very much a minority view amongst most environmentalists, as reflected by global organisations such as Greenpeace, WWF, Friends of the Earth and so on, although there is a small group of dissenters/defectors, much lauded by the nuclear lobby. However, the Vision poll suggests that many scientists, including climate scientists, are pro-nuclear. Whether that is actually the case is unclear. Given its low population (most polls use large samples of 1000 plus) and the fact that it was by open invitation, it is hard to know if the Vision poll is representative. Does it reflect overall global scientific opinion or mainly that in the US and UK? And what sort of scientists? Moreover, it is possible for lobby groups (on either side) to high-jack open polls like this (making sure members and colleagues enter votes), so we can end up mainly measuring lobbying strength and vested interests.

Certainly there are many scientists, especially in the physical sciences, who are pro-nuclear: to an extent that may fit with their conception of scientific progress. However, there may also be a more direct reason for support amongst scientists and engineers directly involved with nuclear power, whose careers and jobs depend on it. Of late, after a lean period, there have been some quite large funding allocations for new nuclear work. Some of this has involved research on materials, which is indeed vital if new high temperature, high radiation flux fission and fusion reactors are to be developed. In the UK, the 'Keep the Nuclear Option Open' programme provided support for continuing and expanded research, and support has more recently been enhanced with £20 million for new nuclear R&D over four years (BEIS 2016b). In addition to some new postgraduate university courses, a National College for Nuclear is also being set up, aiming to train 7000 people by 2020. In the US, on a larger scale, the Obama administration's budget for fiscal year 2016 included over $900 million for developing new nuclear technologies, on top of the private–public 'Gateway for Accelerated Innovation in Nuclear' programme (White House 2015).

Given funding like this, and with major capital investment projects also proposed in some countries, it is not surprising that support for nuclear is high within parts of the scientific and engineering community and related industrial sectors. It is still a major enterprise. A recent global survey of 583 energy executives/experts by the Lloyd's Register included 154 from the nuclear sector, the report noting that this sector was, unsurprisingly, *'united in its belief that it will play a key role in the low carbon landscape'*. However, unhelpfully, the views on nuclear of the other respondents, 323 of whom worked in the renewables sector, were not reported, although they may well not be quite so positive (Lloyd's Register 2017).

Certainly there is much professional and academic opposition to nuclear from outside the industry, as reflected in many publications and in the membership of the UK's Nuclear Consult expert group, the Scientists for Global Responsibility organisation and the work of the Union of Concerned Scientists in the US. In addition, support for renewable energy is very strong amongst scientists and engineers, as well as the public and environmental groups. The pressure for change is being enhanced by the economic problems the nuclear industry is facing, which seem to be worsening (Smith 2017, Brown 2017), and by the rapid spread of renewables, driven by continuing price falls (Cleantechnica 2017), but it is wider than that. It also involves the emergence of a new perspective on how we should go about defining and meeting our energy needs and responding to climate change and air pollution (Elliott 2015).

Clearly, if priorities do change, and the nuclear drive is reduced or halted, major adjustments would have to be made. However, there is no shortage of work to be done on the many non-nuclear energy options. There are already over 8 million people employed in renewables globally, even leaving out large hydro, with job opportunities expanding, by around 5% pa, more in some countries, as renewables continue to grow (REN21 2016). In 2016, over seven times more people were employed in electric power renewables generation in the US (near 500 000 in all) than worked in the nuclear power sector (just over 68 000). The latter was around 21% more than in 2015, but the very much larger solar workforce had increased by 32%, while wind employment, also already larger than in nuclear, rose by 25% (DoE 2017).

Perhaps as the next part of the nuclear and renewables story unfolds, views will change amongst industry insiders, as well as decision makers. At the launch of the 2015 edition of the *World Nuclear Industry Status Report,* lead author Mycle Schneider said: *'The gap between the perception of the nuclear sector by decision-makers, the media and the public and the general declining trend as well as the deep crisis that threatens the very existence of some of the largest players is puzzling. A thorough reality check is urgently needed, especially in countries like the U.K.'* Hopefully this book will play a role in that process.

Further reading/videos

There are many texts, reports and websites on the history and prospects of nuclear power. The present text focusses mainly on US developments, since work on nuclear technology there tended to be more prolific, and it provides many references, including to some historical reviews. For a recent listing and short review of current US advanced nuclear Generation IV projects see: http://www.thirdway.org/report/the-advanced-nuclear-industry.

For UK historical coverage, C N Hill's 2013 book An Atomic Empire: *A technical history of the rise and fall of the British Atomic Energy programme* (Imperial College press) is fascinating in its detail. Simon Taylor's 2016 *The Fall and Rise of Nuclear Power in Britain: A history* (UIT Cambridge) brings the story up to date in a lively fashion, focusing on the political and policy issues.

Bill Nuttall's 2005 book *Nuclear Renaissance* (IoP), though now dated and adopting a different stance to the present study, is still a useful guide to the options and issues. The compilation of mostly critical policy views that I edited, *Nuclear or Not* (Palgrave, 2010) may still also be of interest. Walt Patterson's seminal book *Nuclear Power*, first published in 1976, but still a mine of information on the history and related issues, is now available free at: http://www.waltpatterson.org/npcover.htm

For current and ongoing developments, see the World Nuclear Association's extensive pro-nuclear coverage: http://world-nuclear-news.org/ For a more critical approach, with some nuclear coverage, but with the focus mainly on renewables, see the links to my regular weekly, monthly and bimonthly blogs at: https://renewnatta.wordpress.com For critical views and news on nuclear globally, see the WISE website: https://www.wiseinternational.org/ and its US counterpart NIRS: https://www.nirs.org For detailed regular coverage of UK nuclear issues see NuClear News: www.no2nuclearpower.org.uk/ or from a pro-nuclear perspective: http://www.sone.org.uk and http://newnuclearwatch.eu

The history of the early days lends itself to videos, and there are many on *You Tube*. Some are noted in the text. They include footage from the time, as does an interesting documentary on the USA's early 1950s Experimental Breeder Reactor 1: https://www.youtube.com/watch?v=KncekYvqyWs More generally see the spirited account of the choice between plutonium breeders and molten salt thorium reactors at: https://www.youtube.com/watch?v=bbyr7jZOllI And for the full Windscale 1957 fire story: https://www.youtube.com/watch?v=ElotW9oKv1s&list=PLF193E10985162BA

References

ACS 2012 *Nuclear power plants can produce hydrogen to fuel the "hydrogen economy"* American Chemical Society news item, March 25th http://www.acs.org/content/acs/en/pressroom/newsreleases/2012/march/nuclear-power-plants-can-produce-hydrogen-to-fuel-the-hydrogen-economy.html

Ashley S, Fenner R, Nuttall W and Parks G 2015 Life-cycle impacts from novel thorium–uranium-fuelled nuclear energy systems *Energy Conversion and Management* **101** 136–50 http://www.sciencedirect.com/science/article/pii/S0196890415003957

BEIS 2016a *Energy and Climate Change Public Attitude Tracker* Wave 19, Department for Business, Energy & Industrial Strategy, London, October http://www.gov.uk/government/statistics/public-attitudes-tracking-survey-wave-19

BEIS 2016b *Funding for Nuclear Innovation* Department for Business, Energy & Industrial Strategy, London https://www.gov.uk/guidance/funding-for-nuclear-innovation

Blowers A 2016 *The Legacy of Nuclear Power* (London: Taylor and Francis)

Boccard N 2014 The cost of nuclear electricity: France after Fukushima *Energy Policy* **66** 450–61 http://www.sciencedirect.com/science/journal/03014215/66/supp/C

Brown P 2017 Time and money run out for nuclear revival *Climate News Network* January 11th http://climatenewsnetwork.net/time-money-run-nuclear-revival/

Cany C, Mansilla C, Da Costa P, Mathonnière G and Thomas J-B 2015 Nuclear Power: a promising backup option to promote renewable penetration in the French power system? ed A Sayigh *Renewable Energy in the Service of Mankind* Vol II (Berlin: Springer) (also Heidelberg, London, New York *et al*) pp 69–80 http://link.springer.com/chapter/10.1007/978-3-319-18215-5_7

Carrington D 2016 *After 60 years, is nuclear fusion finally poised to deliver* The Guardian, December 2nd http://www.theguardian.com/environment/2016/dec/02/after-60-years-is-nuclear-fusion-finally-poised-to-deliver

CERRIE 2004 *Report of the Committee Examining Radiation Risks of Internal Emitters* Committee report for the Government, London http://www.rachel.org/lib/cerrie_report.041015.pdf

Clarke A 2016 *Rethinking the Environmental Impacts of Renewable Energy: mitigation and management* (London: Routledge)

Cleantechnica 2017 Cost reduction data from wind and PV solar, Cleantechnica https://c1cleantechnicacom-wpengine.netdna-ssl.com/files/2016/12/1-75pCVZqFXs0HehgOjdbQvQ.png

Cochran T, Feiveson H, Patterson W, Pshakin G, Ramana M, Schneider M, Suzuki T and von Hippel F 2010 *Breeder Reactor Programs: History and Status* Report 8, International Panel on Fissile Materials, Princeton http://fissilematerials.org/library/rr08.pdf

Delucchi M and Jacobson M 2013 Meeting the world's energy needs entirely with wind, water, and solar power *Bulletin of the Atomic Scientists* **69** http://bos.sagepub.com/content/69/4/30.full

Diesendorf M 2016 Renewable energy versus nuclear: dispelling the myths *The Ecologist* April 19th: http://www.theecologist.org/News/news_analysis/2987577/renewable_energy_versus_nuclear_dispelling_the_myths.html

DoE 2006 *Department of Energy Announces New Nuclear Initiative* US Dept. of Energy, Washington DC, February 6th http://energy.gov/articles/department-energy-announces-new-nuclear-initiative

DoE 2017 US energy and employment report *US Dept. of Energy* Washington DC https://www.energy.gov/sites/prod/files/2017/01/f34/2017%20US%20Energy%20and%20Jobs%20Report_0.pdf

Economist 2016 'Hinkley Pointless', editorial leader in the economist August 4th http://www.economist.com/news/leaders/21703367-britain-should-cancel-its-nuclear-white-elephant-and-spend-billions-making-renewables

Elliott D 2013 *Renewables: a review of sustainable energy supply options* (Bristol: Institute of Physics Publishing) http://iopsceince.iop.org/book/978-0-750-31040-6

Elliott D 2015 *Green Energy Futures* Palgrave-Pivot, Basingstoke http://www.palgrave.com/gb/book/9781137584427

Elliott D 2016 Balancing green power (Bristol: Institute of Physics Publishing) http://iopscience.iop.org/book/978-0-7503-1230-1

Emanuel K 2013 Text of the Open Letter from K Caldiera, K Emanuel, J Hanson and T Wigley https://plus.google.com/104173268819779064135/posts/Vs6Csiv1xYr

EP 2016 Clean Energy Emergency, Environmental Progress lobby group, November1st: http://www.environmentalprogress.org/clean-energy-crisis

ERW 2014 *Climate-change investment should focus on retiring coal* IoP/Environmental Research Web, July 23rd http://environmentalresearchweb.org/cws/article/opinion/57973

Flocard H 2015 *Molten Salt Fast Reactor Technology - An Overview'*, Energy Matters website July 20th http://euanmearns.com/molten-salt-fast-reactor-technology-an-overview

Gilbert A, Sovacool B, Johnstone P and Stirling A 2016 Cost overruns and financial risk in the construction of nuclear power reactors: a critical appraisal *Energy Policy* on line April 13th http://www.sciencedirect.com/science/article/pii/S0301421516301549?np=y

Green J 2014 Can PRISM solve the UK's plutonium problem? *The Ecologist* February 26th http://www.theecologist.org/News/news_analysis/2297881/can_prism_solve_the_uks_Plutonium_problem.html

Harvey D 2010 *Carbon Free Energy Supply* (London: Earthscan)

Hirsch R 2015 Fusion research: time to set a new path *Issues in Science and Technology* **31** http://issues.org/31-4/fusion-research-time-to-set-a-new-path/

Hirsch R 2016 Revamping fusion research *Journal of Fusion Energy* **35** 135–41 http://link.springer.com/article/10.1007/s10894-015-0053-y

HMG 2013 *Long-term Nuclear Energy Strategy* HM Government, London http://www.gov.uk/government/uploads/system/uploads/

IEA 2015 *Southeast Asia Energy Outlook 2015* International Energy Agency Paris http://www.iea.org/publications/freepublications/publication/WEO2015_SouthEastAsia.pdf

IEA 2016 Energy, climate change and environment insights *International Energy Agency* Paris http://www.iea.org/publications/freepublications/publication/ECCE2016.pdf

IRNS 2016 Considerations on the performance and reliability of passive safety systems for nuclear reactors *IRSN news release* January http://www.irsn.fr/EN/newsroom/News/Documents/IRSN_Passive-safety-systems-for-nuclear-reactors_01-2016.pdf

IRSN 2015 *Generation IV nuclear energy systems: IRSN presents an overview of the "safety potential" of the systems studied in the context of the Generation IV International Forum*, IRSN website, April 27th http://www.irsn.fr/EN/newsroom/News/Pages/20150427_Generation-IV-nuclear-energy-systems-safety-potential-overview.aspx

Kharecha P and Hansen J 2013 Prevented mortality and greenhouse gas emissions from historical and projected nuclear power *Environ. Sci. Technol.* **47** 4889–95 http://pubs.giss.nasa.gov/abs/kh05000e.html

Kidd S 2015 *Nuclear myths - is the industry also guilty*? Nuclear Engineering International, June 11th http://www.neimagazine.com/opinion/opinionnuclear-myths-is-the-industry-also-guilty-4598343

Koomey J, Hultman N and Grubler A 2016 A reply to "Historical construction costs of global nuclear power reactors" *Energy Policy* online April 13th http://www.sciencedirect.com/science/article/pii/S0301421516301549

Leibreich M 2016 Bloomberg New Energy Finance Summit presentation, April 5th http://www.bbhub.io/bnef/sites/4/2016/04/BNEF-Summit-Keynote-2016.pdf

Lloyd's Register 2017 Technology Radar reviews: 'A Nuclear Perspective', and 'Low Carbon report', Lloyd Register, London www.lr.org/en/low-carbon-power/technology-radar.aspx

Lockheed Martin 2016 Compact Fusion report on company website http://www.lockheedmartin.com/us/products/compact-fusion.html

Lovering J, Yip A and Nordhaus T 2016 Historical construction costs of global nuclear power reactors' *Energy Policy* **91** 371–82 www.sciencedirect.com/science/article/pii/S0301421516300106

Lovins A 2015 The nuclear distraction *Bulletin of the Atomic Scientists* December 18th http://thebulletin.org/commentary/nuclear-distraction

MacKay D 2009 *Sustainable Energy - without the hot air* self-published/UIT Cambridge http://www.withouthotair.com/

Morison R and Shankleman J 2016 Nuclear-Free Zone: Britain Weighs Energy Options Without Hinkley Bloomberg LP, August 1st https://www.bloomberg.com/news/articles/2016-08-01/nuclear-free-zone-britain-weighs-energy-options-without-hinkley

Marshall W 1987Comments as Chair of the CEGB, in a TV interview, quoted in a Parliamentary debate on 30th October 1987 https://www.theyworkforyou.com/debates/?id=1987-10-30a.585.1

McCombie C and Jefferson M 2016 Renewable and nuclear electricity: comparison of environmental impacts *Energy Policy* **96** 758–69 http://www.sciencedirect.com/science/article/pii/S0301421516301240

Mecklin J (ed) 2017 Should nuclear power be a major part of the world's response to climate change *Bulletin of the Atomic Scientists Special Issue* 73 http://www.tandfonline.com/toc/rbul20/73/1?nav=tocList

MIT 2009 *The Future of Nuclear Power' update to the 2003 MIT report Massachusetts Institute of Technology* (Cambridge: Cambridge University Press)

NDA 2016 *Societal aspects of geological disposal* UK Nuclear Decommissioning Authority, RWM report, April 13th https://rwm.nda.gov.uk/publication/societal-aspects-of-geological-disposal/

Neutron Bytes 2016 *Book Notes: Is Anti-nuclear Advocacy a Threat to the Planet?* July 25th https://neutronbytes.com/2016/07/25/books-notes-is-anti-nuclear-advocacy-a-threat-to-the-planet/

Nuttall W 2005 *Nuclear Renaissance: Technologies and Policies for the Future of Nuclear Power: Technologies and Policies from the Future of Nuclear Power* (Bristol: Institute of Physics Publishing)

Patterson 2015 *Electricity v Fire: the fight for our future* self published via Amazon www.amazon.co.uk/dp/B00W5HO1RY

Payton M 2016 *Nearly 50 countries vow to use 100% renewable energy by 2050*, The Independent, November 18th http://www.independent.co.uk/news/world/renewable-energy-target-climate-united-nations-climate-change-vulnerable-nations-ethiopia-a7425411.html

Populus 2016 Hinkley Point Survey, Populus, September http://www.populus.co.uk/wp-content/uploads/2016/09/OmHinkley_Point-wave-2.pdf

Porritt J 2015 Foreword in Schneider M and Froggatt A *et al* 'World Nuclear Industry Status Report 2015 http://www.worldnuclearreport.org/-2015-.html

Raugei M 2013 Comments on 'Energy intensities, EROIs (energy returned on invested), and energy payback times of electricity generating power plants'—Making clear of quite some confusion *Energy* **59** 781–2 http://www.sciencedirect.com/science/article/pii/S0360544213006373

Raugei M, Carbajales-Dale M, Barnhart C and Fthenakis V 2015 Rebuttal: 'Comments on "Energy intensities, EROIs (energy returned on invested), and energy payback times of electricity generating power plants"—making clear of quite some confusion' *Energy* **82** 1088–91 http://www.sciencedirect.com/science/article/pii/S0360544214014327

REN21 2016 *Renewables 2016 Global Status Report*, Renewable Energy Network for the 21st century http://www.ren21.net/status-of-renewables/global-status-report/

Roche P 2016 AP1000 reactor design is dangerous and not fit for purpose *The Ecologist*, November 21st: http://www.theecologist.org/blogs_and_comments/commentators/2988356/ap1000_reactor_design_is_dangerous_and_not_fit_for_purpose.html

Rosen A 2016 *Why nuclear energy is not an answer to global warming* International Physicians for the Prevention of Nuclear War, Germany, London Medact Conference submission, December http://www.nuclearpolicy.info/wp/wp-content/uploads/2017/01/Why-nuclear-energy-is-not-an-answer-to-global-warming.pdf

Smith N 2017 *The Dream of Cheap Nuclear Power Is Over*, Bloomberg View, January 31st https://www.bloomberg.com/view/articles/2017-01-31/the-dream-of-cheap-nuclear-power-is-over

Squassoni S 2017 The incredible shrinking nuclear offset to climate change *Bulletin of the Atomic Scientists* January 4th http://pulitzercenter.org/reporting/incredible-shrinking-nuclear-offset-climate-change

Strauss L 1954 Speech to the national association of science writers (NRC press release) https://www.nrc.gov/docs/ML1613/ML16131A120.pdf

Taylor S 2016 *The Fall and Rise of Nuclear Power in Britain: A history* (UIT Cambridge) http://www.uit.co.uk/the-fall-and-rise-of-nuclear-power-in-britain

Temple J 2017 *Nuclear Energy Startup Transatomic Backtracks on Key Promises* MIT Technology Review, February 24th https://www.technologyreview.com/s/603731/nuclear-energy-startup-transatomic-backtracks-on-key-promises/

Terrell A and Dawson A 2016 *UK Electricity 2050 Part 2: A High Nuclear Model* Energy Matters website, October 31st http://euanmearns.com/uk-electricity-2050-part-2-a-high-nuclear-model/

Tyler C 2016 *Small fusion could be huge* Los Alamos National Labs, News item, July http://www.lanl.gov/discover/publications/1663/2016-july/_assets/docs/1663_JULY-2016-Small-Fusion-Could-Be-Huge.pdf

WEC 2016 *World Energy Scenarios 2016* World Energy Council, London http://www.worldenergy.org/wp-content/uploads/2016/10/World-Energy-Scenarios-2016_Executive-Summary.pdf

Weinberg 2016 *Energy Policies Betray Future Generations* Weinberg Foundation website, November 2nd: http://www.the-weinberg-foundation.org/2016/11/02/energy-policies-betray-future-generations/

Weißbach D, Ruprecht G, Huke A, Czerski K, Gottlieb S and Hussein A 2014 Reply on 'Comments on "Energy intensities, EROIs (energy returned on invested), and energy payback times of electricity generating power plants"—making clear of quite some confusion' *Energy* **68** 1004–6 http://www.sciencedirect.com/science/article/pii/S0360544214001601

Weißbach D, Ruprecht G, Huke A, Czerski K, Gottlieb S and Hussein A 2013 Energy intensities, EROIs (energy returned on invested), and energy payback times of electricity generating power plants *Energy* **52** 210–21 http://www.sciencedirect.com/science/article/pii/S0360544213000492#aff4

Wellock T 2016 *Too Cheap to Meter: History of the Phrase*, US Nuclear Regulatory Commission website https://public-blog.nrc-gateway.gov/2016/06/03/too-cheap-to-meter-a-history-of-the-phrase/

White House 2015 *Obama Administration Announces Actions to Ensure that Nuclear Energy Remains a Vibrant Component of the United States* Clean Energy Strategy, Office of the Press Secretary, November 6th https://www.whitehouse.gov/the-press-office/2015/11/06/fact-sheet-obama-administration-announces-actions-ensure-nuclear-energy

WISE 2015 *Can Nuclear Power slow down Climate Change?* WISE International Report, by Jan Willem Storm van Leeuwen, World Information Service on Energy, Amsterdam https://www.wiseinternational.org/sites/default/files/u93/F4%201nuclGHGshare-ED.pdf

WNA 2016 *Nuclear Process Heat for Industry* World Nuclear Association Information Library http://www.world-nuclear.org/information-library/non-power-nuclear-applications/industry/nuclear-process-heat-for-industry.aspx

WNN 2016a *Carbon pricing not enough to help nuclear power* World Nuclear News, June 8th http://www.world-nuclear-news.org/V-Carbon-pricing-not-enough-to-help-nuclear-power-10061601.htm

WNN 2016b *Achieving 1000 GWe of new capacity by 2050* World Nuclear News coverage of the World Nuclear Association 2016 annual symposium, September 16th http://world-nuclear-news.org/NP-Achieving-1000-GWe-of-new-capacity-by-2050-16091601.html

WNN 2017 *US consortium calls for public-private SMR support* World Nuclear News, February 20th http://www.world-nuclear-news.org/NN-US-consortium-calls-for-public-private-SMR-support-2002177.html

Wolff G and Jones M 2016 Non-nuclear scenarios listings and links http://www.mng.org.uk/gh/scenarios.htm

WWF 2013 *100% Renewable Energy by 2050 for India* World Wildlife Fund for Nature with TERI, New Delhi http://www.wwfindia.org/?10261/100-Renewable-Energy-by-2050-for-India

WWF 2014 *China's Future Generation* World Wildlife Fund for Nature http://worldwildlife.org/publications/china-s-future-generation-assessing-the-maximum-potential-for-renewable-power-sources-in-china-to-2050

WWF 2016 *Vietnamese power sector can reach 100% renewable energy by 2050, according to new study* WWF Global http://wwf.panda.org/wwf_news/?267471/new-study-vietnam-power-sector-and-renewable-energy-by-2050

IOP Concise Physics

Nuclear Power
Past, present and future
David Elliott

Appendix

Nuclear and renewables—the basics compared

Nuclear fission and fusion both create large amounts of heat from small amounts of material, some of which is relatively rare. Complex technology is needed to initiate and control the reactions. We have learnt to do this for fission reasonably well, but not yet for fusion. Converting heat to electricity is a more conventional problem, and we have, over the centuries, got good at doing it, by operating at high temperatures and pressures. However, the (steam-based) process is thermodynamically limited, to around 30%–35% efficiency, although the overall energy conversion efficiency can be increased by the use of some of the waste heat.

Renewable sources like wind, wave, hydro, tidal and solar, are (unlike fissile fuel sources) unlimited in resource terms, but are diffuse, although, in most cases, widely distributed. We have learnt how to extract the energy in these natural energy flows and convert it into electricity directly, with conversion efficiencies of 60%–70%, or more in some cases, but lower in others (e.g. around 20%, but improving, for PV solar). Relatively large catchment areas are needed e.g. for wind or solar farms. However, in the case of solar, existing roof-tops can be used, so can desert areas, and in the case of wind, the area of the turbine bases is only a small fraction of the area of the wind farm: most of the land can be used for other purposes e.g. crops or grazing. With offshore wind (and wave and tidal stream turbines), no land is used. The use of biomass inevitably requires more land/kWh than other renewables, basically due to the low efficiency of photosynthesis, but unlike wind, solar etc, it is a storable fuel.

Extracting energy from relatively low energy density flows does require space, but given the low diffuse flows, the extraction process usually only results in relatively minor environmental impacts and risks, depending on how much energy is extracted. By contrast, nuclear power involves the use of concentrated energy sources and very high energy densities processed in relatively small plants. That is part of the attraction. However, the down-side is the risk of major accidents and impacts, if control of the intense energy conversion processes fails, with the release of dangerous materials. Dangerous long-lived wastes are also inevitably produced as part of the conversion process and have to be isolated from contact with the ecosystem, water courses included, for many centuries ahead.

Technological improvements can help reduce some of these nuclear risks, e.g. by improving fuel use efficiency so less waste is produced, but they cannot be entirely avoided. They are part of the nature of nuclear power and pushing technology and material science to the limit can be risky. Renewables will always be diffuse, but technical improvements can raise energy conversion efficiency and reduce costs and land use, without increasing the risks. To put it simply, it is, arguably, a gentler path ahead.

The portrayal of renewables as inherently 'gentler' may of course been seen as overblown. Large hydro plants can have major impacts, and all technologies have some impacts. However, it can be argued that creating intense energy and radiation fluxes and long-lived wastes via nuclear fission moves us into a new league, with the resultant impacts, risks and side effects taking us beyond what is acceptable. The implication is that we should abandon this approach. Some might also say the same of fusion. That may be too extreme. Some would say that we should persevere in trying to apply our hard-won understanding of sub-nuclear physics to energy production. It may be that our initial approach, fission, has problems, but the wider enterprise should continue.

It is an interesting debate, at base, about, on one hand, the limits and potential of science, and on the other, our real world experience of the results of its application and the limits to our engineering prowess. Can we do better? In his wide ranging self-published book *Electricity v Fire* Walt Patterson focuses mainly on fossil fuel combustion for electricity generation, and argues that it is now clear that we should use it less, use it better, and pay more attention to what we need rather than just rushing to use either fire or electricity. The same might be said of nuclear electricity, although some might wish to see its use abandoned completely, at least on Earth. The debate continues.